THE SHIFT O. ....
THROUGH 2012
AND BEYOND

THE BIGGEST CHANGE
CHALLENGE OF OUR TIME

Bridging Mayan Cosmology with the Road to the
New Earth Beyond 2012

Peter Christian

Acknowledgements

There are a number of people to whom I owe much love and gratitude for being able to write this book. In no particular preference or order, I would particularly like to thank Judy Coleman for much informative inspiration, my dear friend Torben E. for being there through my personal process, my mother and father with whom I have shared a spiritual bond and the many extraordinary teachers, I have had in life. I would also like to thank Ian Henderson for bringing the Mayan Calendar to my attention and P. Krishnamurthy for her moral and other support during the time of writing. The list is long and could go on....

There are many of you trying to bring the world forward by simply being who you are and doing what you know best. So, I will stop here and express my respect for sharing of your love, time and gifts wherever you are. The universe knows who we are and one day humanity will remember or pay tribute to the mark we came here to make at this pivotal point in the planet's history - hopefully right on time....

Contents

Preface/introduction 1

1 On the Cusp of A Great Cycle 7

2 Entering Our Galactic Day 34

3 Ending Grand Cycles of Time 59

4 The Transformation of Our World 85

5 Reviewing Self 106

6 Navigating Through the Transition 121

7 Finding the Road Ahead 143

8 Journey to Wholeness and Joy 158

9 A Higher Working Consciousness 173

10 The Cosmic and Co-Creative Self 183

References 192

Preface/Introduction

Many people have heard of the Mayan Calendar, but know little or nothing of what the Mayan Prophecy and 2012 phenomenon is really all about. Although it is not a fatal prophecy about an apocalypse, it is a crucial time for our humanity and world. The Mayan calendars have predicted that a great cycle of time is ending and as we move into a new cycle through and beyond 2012, we find ourselves in the midst of an unprecedented time of radical transformation and global change – called the Shift of the Ages.

During these years we will be going through a grandiose cosmic event - a crossroads, tipping or choice point which has been foretold by many scriptures, ancient and indigenous peoples. We cannot escape the event, but we can choose how we will think, act and respond to what has arrived right on (cosmic) time to be respected and celebrated. It is a time for us to "grow up", and let go of the catastro-phobia or fears of our so-called humanness and become more of the truth of who we really are....

It is both an individual and shared collective event that concerns our future destinies and relates to the ways we think of and conduct ourselves as human beings in the world. As with any great time of personal or global challenge or change, we are given the opportunity now to realize our true self and potential. It is both a time of emergency and emergence in consciousness - where long held or established human beliefs, negativity, fear or darkness for a while might be prevailing throughout in certain circumstances before something new, greater or better can emerge through possible shocking or explosive events. This is not to scare you. On the contrary it is to prepare you by making you aware that what could become one of our most challenging times or darkest hours – by threatening our human existence through earth quakes, climate change, societal, economical or social upheaval or even war – has arrived right on our doorstep, so that what needs to surface will have to be addressed and transformed into a new or higher order of the day for our human civilization. As you can see when you watch the news, much of this is already happening.

As we move beyond the old cycle that has followed us for eons, and push through the threshold of 2012 and beyond, many of our attempts to hold onto the old world paradigm or era out of fear, control, lack, greed or the like of the ego, which prevent our true selves, connection to God or divinity to emerge, could or will be revoked. This is part of releasing outdated ways of thinking and being that might no longer serve us as we come into alignment with a new cycle or age for the Earth.

Whether we perceive it or not in this way, the 2012 prophecy about a Shift is here as the world is under extreme pressure to change or transform in some way. It is manifesting through the many environmental, financial and cultural crises occurring right now. In times like this where real change is mostly needed, change often causes resistance – so, we have to find the necessary courage, responsibility and humility that will enable us to create a new reality based more on love, truth, respect and peace.

What then appear as two opposing paradigms - or waves of possible light or dark opposing each other - are actually part of the same (energy) being played out in different ways or forms from different perspectives according to our individual choices, consciousness and shared (soul) experience. If we can trust and allow what is meant to happen to occur while opening ourselves to a higher vision, fuller sensory or spiritual awareness, the Shift can start to work for and with us to make the changes or adjustments we have to make. It is a matter of going with the flow rather than trying to resist it. What initially may seem chaotic or causing us much confusion or many problems is actually a new wave which is guiding us through a spiritual growth experience.

Since the beginning of mankind we have been wondering about our true spiritual origin and the nature of our human existence. These questions are brought to the fore around the Shift to stretch our perceptions and possibilities beyond the Mayan Calendar end date itself. December 2012 is only a small, yet significant signpost or open window in time as we will see, which allows us to "shift" with implications for the future decades and even generations ahead. If we take the hints from the Universe, we are going to exist and even prosper in the post-2012 world or era, but what if we don't?

This is why I wrote this book, and why we need to get our act together now to be able to put ourselves in a position where we cannot only deal with these years, but also be able to shape the future ahead. We cannot afford to sit around and wait - because time is accelerating and the future will soon catch up with us.

At the time of writing this book, many unexpected world events have been taking place that seemed impossible just a few years ago. The further we go along the trajectory of the Shift, the more events seem to be happening at an increasing rate. The question is how capable, ready or adaptable we are?

In order to cope, the first thing to do is to try to understand what this Great Shift of the Ages is really all about and that we are not separate from the universe or planet of which our world is a part. If we were able to look back at this time period from a point in the future or through the eyes of our grand children much would be seen about how much we had experienced or how far we had travelled.....

Initially, it might seem difficult to grasp some of the spiritual ideas or concepts which are laid out in this book, but essentially this is what the Shift is about: A rare spiritual event and process which outlines new ways of thinking or being for us! If you skim through the book – or stop in the first half, you might miss important parts, but if you read carefully or listen to its message between all the pages, you will see how the different pieces of the "Shift of the Ages through 2012 and Beyond" fit together in a grander puzzle and can even serve as a personal guide for the years ahead. There are many things we need to sort out or have to process through personal experience as we approach the future ahead. This is why the second half of the book might make more sense later in time. We all have to pass through this Shift in different ways on different trajectories depending on who we are. Gradually this will all start to make more sense...

At the time when the book is getting ready, we are at the 7th day of the Mayan Calendar and the Shift is dramatically picking up speed with many new codes of light flowing into our planet as the 9th wave begins. As it meets with old ideas and concepts the change waves of the Shift will rapidly increase. How the Shift and the world moves on after the proposed end date of the Mayan Long Count Calendar at the winter solstice, December 2012, I believe, is far from set in stone, but essentially, and perhaps perfectly so, entirely up to us and our consciousness. I hope this book will inspire you to make the best and most of it.

Peter Christian,

Psychologist, journalist, teacher

October, 2010

"Without darkness nothing comes to birth,

as without light nothing flowers"

- Anonymous

PART 1

THE MAYAN END AGE PROPHECY

Photo (previous page)

Mayan Calendar wheel crop circle, Silbury, England, 2004

Image credit: Lucy Pringle

CHAPTER 1

ON THE CUSP OF A GREAT CYCLE

What Is the 2012 End Date About?

We are living right now on the cusp of the Mayan end age, which marks the closing of several cycles of time. The first is the end of the Mayan Calendar galactic time period - a "precession" of the equinoxes, the "Annus Magnus" or "Great Year". This Great Year of 25,625 years is divided into five cycles of 5,125 years. We are almost at the end of the final 5,125 year cycle which ends on 12-21-2012. Some say it also signifies the end of an even greater cycle of 104.000 years related to the Pleiades Star System.

Why Are We Talking About the 2012 Phenomenon?

Although the winter solstice on December 21, 2012, precisely at 11:11 a.m. Universal Time – according to many mathematically inclined scholars on the subject - marks the completion of the 5,125 year Great Cycle of the Ancient Maya Long Count Calendar, contrary to what many believe, it is not a simple, linear end-point in time.

Following Mayan concepts of cyclic time and World Age transitions, it was considered by the ancient Maya to signify the completion of a World Age Cycle or creation of a New World Age.

There are many records about the end of vast cycles of time. In recent years the "2012" Hollywood movie has also served to alert millions of people of the 2012 phenomenon, in case we had not already heard the "bad" news. The film exposed our human vulnerabilities and fears of the unknown with its global disaster scenarios being the end of the world without explaining what 2012 is really all about; the closing of a World Age Cycle.

The Mayan prophecy is really about themes of planetary transformation and renewal in the history of our planet. Being the end of a huge cosmic cycle - spanning approximately 26,000 years - it is a moment in time which has never existed before in modern human history. The question is will we respond to it? How?

Mayan timekeepers believe that human evolution unfolds as a result of such precisely calibrated master cycles of time. They predict that Earth and humanity are about to be birthed into a new reality predicated to cause a dramatic advance in consciousness.

Instead of being something to fear, rather we can understand the 2012 prophecy signalling to us that we need to awaken and realize that we are living in pivotal, landmark times. What this new cycle in time - over a period - has in store for us is yet somewhat of a mystery that will be revealed and experienced as it unfolds.

In the meantime, our human population is growing at an accelerated pace while our natural environment and marine eco-systems are stretched to the limits. We rely as peoples worldwide on them to survive. The old world's ignorant mentality, focused on greed, endless economic growth and unconscious consumer materialism in the West, which has now been adopted by the East, has reached dangerous peaks, is irresponsible and cannot continue for another 75-150 years without life on our planet collapsing.

In this way we are already at a collective cross-road, a moment in time that is calling out to us to participate or get involved in new (more sustainable) ways. Some say we are being summoned to use our true capacity as human beings to bring inspiration, leadership and empowerment to our human civilization while reaching out to others and Mother Nature as well. The necessity and possibility for change is at our fingertips like never before.

While we might be living together in a time of great uncertainty, spiritual initiation or a re-birthing process, while we are going through times of planetary changes or challenges, a new paradigm is trying to emerge in our world, through our hands, hearts and minds. Like a planted seed trying to grow through the cracks of a pavement or sidewalk, new comprehensions and solutions that better honour who we are and what we can do also for our planet are needed on our collective path - from new sciences and economic models, healing modalities and energy technologies to new educational formats and forms of conflict resolution etc.

In this way, we can perceive the 2012 completion as a reminder to us from the Ancients or the universe itself, that right here and now, we are living in daunting and auspicious times. A time of emergency and emergence where we can realize how critical the above is and that we need to start working together in respect for all life, by becoming more conscious of how our intentions and actions affect the Whole and start to do something about it.

This is perhaps the most important theme if we are to become more conscious now of all that has been unconscious in the past. It is time to see and know how we can become aware, so that we can participate directly in this great transformation from the energies we transmit to the details of our lifestyle choices.

It is a time where we are catalyzed to awaken to our personal abilities and collective responsibilities with greater awareness, if we are to learn to unify our purpose with the great planetary equation. The new cycle that we are entering can be founded on our awakening to the beauty and responsibility of this interconnectedness.

If we can connect with each other, we can discover what these mysterious times mean to us and our lives. By recognizing we are in a global time of change, we can simultaneously shake off any wounded victim mentality we might have, and possibly arise as way showers, "warriors" and caretakers of our planet for future generations. It is about growing up as human beings and become aware of the responsibility we have in order to do the necessary ground work to birth a New Earth based on truth - obliged as we are to root new thoughts in our consciousness that others can follow. We can try to brush it off or avoid it, but the more we do, the more we will be wasting precious time.

Are We Approaching the "End of the World?"

No. Not in terms of a complete destruction or Earth annihilation scenario. The years around 2012 are not going to bring the physical "end of the world", but .....When we contemplate the expression "end of the world", we must realize it can mean other things. It can refer to a cycle; a period of time; a world age and with it a way of life. According to Mayan cosmology, 2012 is really signalling the completion of one World Age Cycle which means we are transiting into another, new world, age, era or cycle.

It is said by Mayan elders that the world we are ending is the world as "we have come to know it" - one which has been dominated by the exploration of materialism through the separation and greed of our ego consciousness. Therefore, the world which will follow will be founded on different values that honour the spirit of the interdependence of all life through greater balance.

The term "End of Age" is also linked with "the end of linear time", and of course no one can say for sure how valid these claims are or may be. Many of the Mayan Calendar "end of time" concepts are based more on fear than on what will actually transpire around 2012.

The more conscious we become of the non-linear, synchronistic nature of existence, the more rapidly we may evolve beyond the rational-material linear time paradigm. According to the Mayan Calendar system, that should not be a problem now that our attention is brought to the 2012 phenomenon.

At the time of writing, the living Maya of Mesoamerica urgently want it to be known that their ancient prophecies have been distorted as doomsday predictions by ignorant sensationalized 2012 rumours which are not coming from the Maya themselves, even though it may appear to be associated with them. They are very upset about this.

The ever-increasing fascination with 2012 as the "end of the world" in recent years, have been projected by the media which reflect our collective psyche. There is an idea or a sense of an impending retaliation or doom that might come down from Heaven as an act or will of God to punish us for our sins or misguided human behaviours as it may have done before on our species in a distant past.

While it is clear we are certainly living in times of great uncertainty or imbalance, we need to stop terrorizing ourselves by spreading hysteria or rumours that do not assist us in rising to the challenges, we are facing. Fear is a primordial trait in human nature based on how vulnerable or fragile we are. We need to get our heads out of the ground and understand that if we allow too much fear mongering to enter our collective thoughts, we can easily become shut down, delusional or reactionary in ways that feed the fear rather than motivate us through greater discernment and awareness.

When we contemplate what the world might be or look like beyond 2012, let us be clear that no one can predict the specifics of how things will unfold during such a gigantic cosmic event where we enter a whole new cycle, according to the Mayan Prophecy.

There is not one path or way that will lead or take us there. To navigate these critical times on Earth, we should learn to follow our hearts and inspiration, for they are our purest guides and can help us in tuning a potential chaotic or destructive path into a more positive future of human growth – that will serve us all.

Whether the transformation and changes will come gradually or swiftly, be positive or negative, bring us down, together or tear us apart, depends to a degree or extent on each of us.

Any re-booting of the Mayan Calendar in our times will not so much come from a date on the calendar itself, but from the level of awareness and consciousness which transforms and transcends the present one which humanity has risen up to during these "end times".

This is or can become quite complex for sure, but nobody said it was going to be easy. Many challenging situations may arise - because although we inhabit the same planet, no one can predict which way it will go with our many human disagreements. In many ways we still act as if we are totally different in spite of the easy interconnectedness of our modern high tech world and the fact that we are members of the same species.

What Are Our Indigenous Peoples of the Earth Saying?

The reason why we should pay more attention to our ancient and indigenous peoples around the world at this time is because their spiritual wisdom traditions have an understanding that supersedes the way we hitherto have seen ourselves in relation to the cosmos and the Earth along the way of modern scientific and technological exploration.

Both the Hopis and modern day Mayans recognize that we are approaching the end of a World Age... The Hopi and Mayan elders, however, do not prophesy that everything will come to an end. Rather their messages concern our making a choice of how we enter the future that lies ahead. How we move forward with either resistance or acceptance may determine whether the transition will happen with cataclysmic changes or gradual peace and tranquillity. In Inca language this time is referred to as the time "to merge with the universe". In Andean prophecy, it refers to a new Golden Age of the human experience. It is known as the age of meeting ourselves, and heralds a time where we are coming together again across all nations and peoples. The Australian Aboriginals refer to a Dream Time from which our ancestors come and return.

The same theme of constant renewal and regeneration permeates our future destinies and generations and can be found reflected in the prophecies of many other Native American visionaries from Black Elk to Sun Bear. It is a universal ingredient in all true prophetic pronouncements which reflect the hidden natures and motivations of human behaviour, as well as our future possibilities. True prophecy is more than merely a forecast. Its purpose is to provide the lessons that are needed to be learned from a potential future prognostication so that, if possible, the lesson can be processed beforehand and the course of the future can be altered to manifest a different pathway or a different reality of the prophesied events.

In this book's context, the period of time leading up to the end date in 2012 is the benchmark, and the period after appears to become a very decisive time when important choices will have to be made where any number of future scenarios are possible. The Mayan prophecy about the Shift of the Ages through 2012 is our true guide to determine what those different timelines are and how we can make the right choices for the future. Isn't it about time that we truly listen?

What Is Said About the Shift of the Ages and Why Should We Care?

Consensus indicates that we are coming out of what the Indian tradition names Kali Yuga (the age of darkness/ignorance) and are on the verge of entering the Satya Yuga (the age of truth) when all falsehood will expose itself and drop away. The Yuga which links these two is named Krita Yuga (age of transition).

From the Western astrological perspective this seems to correspond with an understanding that we are transiting from the Piscean Age to the Age of Aquarius. The onset of the Aquarian Age has spoken to us since the late 1960s of spiritual awakenings and changes based on an increased awareness of our spiritual self and social interconnection. It is a time of rebirth, great spiritual development and new scientific discoveries heralding a time of greater possibility and joy.

In the Christian tradition the generally accepted idea is that the New Millennium starts around the year 2000. Webster's 9th Collegiate Dictionary, USA, 1983 gives a definition as follows: - "The 1000 years mentioned in the Book of Revelation, chapter 20 during which Holiness is to prevail and a period of great happiness or human perfection".

From the Book of Prophecies of the Knight John of Jerusalem (11th Century) comes the following: "The millennium that comes after this millennium will change into a light time. People will love and share and dream, and dreams will come true." It adds: "People will be one big body of which every person is a tiny part. Together they will be the heart and they will speak one language. When men will have reached heaven and know the Spirit of all things...People will receive a second birth and the Spirit will come into them."

In the Islamic tradition there are many instances where the Holy Koran and the Hadith mention a future time of judgement & resurrection, known as the Qiyamah time. In the traditions of the prophet, this time is indicated as coming some time after 1400 years (in the Hijri calendar), which seems to coincide with the beginning of the twenty first century.

For Buddhists there is some expectation that the Wheel of Dharma, the metaphorical wheel of time is set to turn for the first time in 2500 years since Lord Buddha's advent. The Buddha apparently taught that each revolution of the wheel signalled a new beginning or rebirth for humanity.

William Blake (1757-1827) speaks of the Judaic tradition in his work: 'The Marriage of Heaven and Hell' V14 as follows: "The ancient tradition that the world will be consumed (overthrown) in fire at the end of six thousand years is true....For the cherub with his flaming sword is hereby commanded to leave guard at (the) tree of life: and when he does, the whole creation will be consumed (overthrown) and appear infinite and holy, whereas it now appears finite and corrupt. If the doors of perception were cleansed every thing would appear to man as it is, infinite. For man has so closed himself up, till he sees all things thro' narrow chinks of his cavern."

So what do these ancient texts really say about the present? What is so intriguing about it is that they all seem to agree in some way - like faint echo from the past which informs us of a hitherto unsuspected phenomenon the coming of a "spiritual personality", which will facilitate a collective rebirth.

The "Mayan Story of Creation" tells about our current time frame where a whole new universal or planetary cycle begins. The significance of this grand opportunity for transformation of humanity's consciousness to occur is shown in pictorial glyphs. Amongst many of them is a final glyph that shows the reawakening of the Spirit of the Feminine, and the transfer of the staff of power from the masculine in this time frame.

Before we can fully comprehend the scale of this, we must try to understand how we have come to know the meaning of the Mayan Calendar itself.

The most important of the books that survived the Spanish armada and conquest of the ancient Maya is called the Dresden Codex, named after the German town library where it is located. This strange book, inscribed with hieroglyphs, was written by Mayan Indians centuries ago. As far back as 1880 a brilliant German scholar read the Dresden Codex and cracked part of the code of the Mayan Calendar, by making it possible to translate some of the dated inscriptions that were found etched on buildings and ancient Mayan artefacts. This translation was concerned with astronomy, providing detailed tables of lunar eclipses and other phenomenon such as the Venus passage which was often associated with war and which we now know coincides with a new passage shortly after 2012.

The Maya Stelae & Codices

Stela 11 from Izapa shows Cosmic Father
(in the "mouth" of Cosmic Mother, the
"dark rift" or "birth canal" in the Milky Way).
This is an image of the celestial alignment
which culminates in A.D. 2012.

Page 9 of the Dresden Codex
(Förstermann edition, 1880).

The Maya codices (singular; codex) are folding books stemming from the pre-Columbian Maya civilization. There are three main Maya codices; the Dresden, Paris and Madrid codex named after the cities in which they eventually settled. The Dresden codex is generally considered the most important because it is the most elaborate of the codices, and also a highly important work of art. Many sections are ritualistic (including so-called 'almanacs'), Others are of an astrological nature (describing eclipses, the Venus cycles etc.)

A Historic Overview of the 12-21-2012 End Date

Although these "end times" have been foretold by many Ancients, the Mayan prophecies are by far the most timely accurate, precise and significant because they are based on real, highly advanced, astronomical data, observation and knowledge that the Maya possessed by studying the movements of the sun, planet(s) and stars – including recent eclipses, centuries ahead of time, which have been confirmed by modern day science. This is quite astonishing, so when, how or where did the Maya develop their timely accurate calendar systems? The history is as important as it is interesting.

The Maya World in Mesoamerica

The Mayan civilization is divided into three time periods which engulfed 3,000 years. The first is the Pre-Classic Period spanning from 2000 B.C.-250 A.D. The second is the Classic Period which spanned from 250 A.D.-900 A.D. The third is the Post-Classic Period which spanned from 900 A.D.-1500 A.D. The Maya lived in the eastern third of Mesoamerica, mainly on the Yucatan Peninsula as a group of related Native American tribes who had the same linguistic organization.

The best known group is the Maya Proper. Who generally occupied the Yucatan. There are other groups as the Huastec, the Tzental; and the Cakchiquel and the Pokomam. With the exception of the Huastec, all of these Mayan groups occupied a continuous landscape and they were all part of the Mayan culture. This culture was the greatest civilization among the original cultures of the New World (western hemisphere). Even though the Mayans had common organization, they were not unified under one empire. As suggested above, there were many separate groups with similar cultural backgrounds. The Mayans had common artistic and religious components, but politically they were mostly independent Mayan states.

The Yucatan Peninsula and the Maya World

The ancient peoples of Mesoamerica were of a vast interconnected empire, filled with rich art, education and final destruction. The Classical Maya were one of these tribes which developed a highly sophisticated civilization in the Yucatan and Guatemala that reached its peak more than 1,000 years ago. Around that time, the great Maya civilization of scientists, artists and warriors, abruptly abandoned their world leaving their cities and temples to be swallowed up by the jungle. They left behind a great legacy of civic centres and temples of worship, houses and apartment buildings, buried great works of art and pottery. The downfall of the Maya is still somewhat unknown. There are several possible theories about their downfall, including soil exhaustion, water loss and erosion, and the competition between agriculture and the Savannah or a dimensional shift. Other possibilities include catastrophes such as earthquakes and hurricanes, disease, abundant amounts of high social structure, wars or invasions by other surrounding people and cultures.

The collapse of the Maya range from the hypotheses stated above to a single catastrophic events or a "dimensional move". With all these possibilities, the collapse of the Maya remains one of the most intriguing events in human history.

Apart from archaeological intrigue about what caused the destruction of their civilisation archaeology also discovered their written language was based on complex pictographs, much like the ancient Egyptian hieroglyphs. Although it has been a monumental task to decipher their meanings, the translations revealed that they were extraordinary architects and astronomers, who were preoccupied with time and developed methods of timekeeping that are far more precise and advanced than our western calendar system. The crowning achievements of their entire civilization, was what has become known as the Mayan calendar system which supersedes any other time system.

The Discovery of the Mayan Calendar System

It is specifically the Mayan Long Count Calendar, ending in 2012, that has sparked great interest, intrigue and debate with researchers, scholars and mystics around the world. Questions remain that still have not been fully answered, such as why does this calendar end and what exactly will occur on the Winter Solstice date of December 21, 2012? Will the calendar reboot itself or is it a complete ending? Whatever we believe in – it is worth taking a look at the discoveries and evidence surrounding the Mayan calendar system first. So, please bear with the explanations....

The Maya developed their calendar with great precision before the Julian or Gregorian calendars came into existence. When the Spanish arrived about 500 years ago they were fascinated and horrified by what they found in this "New World." The Catholic priests looked at the Mayan religion, which included human sacrifice, as barbarous and satanic and set about to destroy it without a trace.

Whole libraries of colourful bark-books were burnt and the Mayans who did not die from disease, hunger or over-work were forcibly converted to Catholicism as they were being forced into submission. One notable figure with these inquisitions was a Spanish monk, Fray Diego de Landa, who was responsible for building consent in Mesoamerica in 1562. He became known to the Mayan people for burning hundreds of their parchment scripts about aspects of Mayan life, customs and beliefs on the request of the Spanish queen..

Fortunately, not all the Spanish were unsympathetic towards the Mayans. A few, such as "friar" Bernadino Sahagun made friends with the natives and attempted to record their traditional beliefs and ideas. A few priceless books and relics survived the destruction, after either having been hidden by the Mayans or exported back to Europe as presents for the Spanish king. While only four of their folding-bark books survived the fanatical purges of the Spanish priests, their writings in stucco, stone and pottery remain. The voices of the ancient Maya stood silent for centuries before the advances in modern archaeology made decipherment possible.

From the 1930s to 1960s the dominant expert in Mayan glyph studies, J. E. Thompson - a British archaeologist - created a meticulous classification system assigned to over 800 Maya signs. After living with the Maya in Southern Mexico and Belize, he concluded that the focus of Mayan Civilisation was time and that their stone tablets were built to commemorate its passage. The figures in their art were priests and Gods and their glyphs were symbols recording the mystery of the Heavens.

A major breakthrough in deciphering and understanding the Mayan Calendar system, was initiated in 1945 when a young Russian army officer found a book reproduction in the ruins of the national library in Berlin of the then known three Mayan codices: The Madrid, The Paris and Dresden Codices. His passion grew when he read an article concluding that Mayan glyphs were decipherable. After attaining a degree in linguistics he took up the challenge and concluded that some of Thompson's ideas where wrong because the Mayan glyphs represented by 800 signs, was rather a logographic system combining both word signs and phonetic signs.

Ancient Maya glyph

One big hurdle then remained in cracking the Mayan code: Namely, reading the hieroglyphs phonetically in the language in which they were written. Although the phonetic decipherment had already begun – fewer than 30 syllabic signs could be read at the time with confidence - as the descendants of the Maya had forgotten the ancient written language when their ancestors' language was subdued by Spanish.

A key to the sound of the glyphs would be found when David Stewart whose educational background as a Mayanist began at an early age when his father often brought his son on field trips to the site of Copal in Mesoamerica. He discovered that the complications of Maya writing are the many substitutions one sound can have with up to 13 -15 different versions of the same glyph as the Maya hated repeating themselves. Understanding these substitutions started to break down the Maya writings and codes into a more manageable, comprehensible system. As the phonetic readings poured out of the temples at Palenque a more complex view of the Maya began to emerge: Instead of being peaceful, stargazing astronomers – the Maya were shown to be much like any other civilization with a great deal of conflict and sacrifice. For the Mayan civilisation the world of the dead and living were intertwined as they considered life and death part of a greater (time) cycle.

For example: One myth painted repeatedly on Mayan vases echoes a 1600 century manuscript from the Copal rule. It tells the tale of two twin brothers summoned to the underworld. They play a ball game against the lords of the underworld in which they are finally defeated and sacrificed. One of the brother's heads is hanged in a tree. A second set of twins are being born by a visitor to the head. They resurrect their father - the maze good by beating the God's of the underworlds and then raise up to become the sun and the moon. It is these myths that have given rise to the universal Mayan creation myth.

Further evidence suggests that in the 9th-10th century power structures or battles began to tear the Maya apart. Different groups where fighting and capturing each other, tearing down and burning cities with many killings and political upheaval – the great cities finally collapsed.

In 1986 Maurice Cotterell put forth a revolutionary theory concerning astrology and sun cycles. He had for some years suspected that the sun's magnetic field had consequence for life on earth. As he studied the Dresden Codex, he discovered that the Mayan Calendar was not arbitrary but might be based on knowledge of sunspots. This explained the Mayan obsession for long cycles of time and their belief in the rise and fall of the four previous ages of man with each emerging new sun cycle. For instance the Dresden Codex depicts a time in the future with heavy flooding or heavy rainfall.

Because Mayan time was cyclical, effects were thought to eventually, potentially repeat them selves. Certain days were considered unlucky, others days were considered good. Not unlike looking at today's news paper horoscopes. To the Mayans certain years were good and certain years were bad. Also certain blocks of 20 years, called katuns were considered good or bad. Each block of time, and its so-called 'personality' for good or bad, feast or famine, were the teeth on the cogs of the calendar wheel.

By studying the heavenly bodies, the sophisticated Mayan civilisation gained knowledge about many cycles which occur on Earth and recognized a 13-day cycle of energy which comes from galactic sources and a 20-day energy cycle coming from the Sun.

By looking at the calendar, the Maya could not only see what day it was, or year, but also what 20 year period it was in relation to. They could often see what was to come. Even the demise of their own civilization may have been foretold as a process of their view of time.

The Mayan collapse is tied into a katun of change and is related to earth changes by earth events, fire or water. The Mayans believed that the world had been destroyed four times before. First, by water, second by wind, third by fire and fourth by earth changes. At the end of each age there is a time of chaos, and then a period of rebuilding as a new age begins. The fifth world is the one we are entering right now – a possible time of more balance and respect for all life, the Earth and our human interconnectedness.

The Mayan Calendar is essentially based on naturally occurring universal cycles with a close connection between the Earth, the Sun, and the stars. Because the Maya honoured Mother Earth and the energy cycles that are present here as well as in the universe, they were in tune with these cycles and wanted to understand them to be able to predict what might occur to their crops, culture and future civilization(s).

By studying the time cycles and their interaction with each other, the Maya gained an in depth understanding of how the evolution of life unfolds on Earth in relation to the cosmos that far exceeds our 'modern' understanding or beliefs. Fortunately, this knowledge has now become available to us through the translation of the calendars.

Understanding the Mayan Calendars

The Maya had many calendars, but the following 3 have been the main focus of modern Mayan Calendar researchers:

The most important of these calendars is the one with a period of 260 days called the Tzolk'in. The Tzolk'in is combined with the 365-day calendar called the Haab to form a synchronized cycle lasting for 52 Haabs, called the Calendar Round. Smaller cycles of 13 days and 20 days were important components of the Tzolk'in and Haab' cycles, respectively in relation to linear time.

A different form of calendar was used to track longer periods of time, and for the inscription of calendar dates. This form, known as the Long Count is based upon the number of elapsed days since a mythological starting-point. According to the correlation between the Long Count and western calendars accepted by the great majority of Maya researchers this starting-point is equivalent to approximately August 11, 3114 BC in the Gregorian calendar.

Their still unsurpassed understanding of numbers and astronomy gave us the Mayan calendars of the Short and Long Counts adding up to longer or grander cycles as the calendars intermeshed with each other - such as the 5,125 year cycles with the latest starting in 3113 B.C. and thus looking toward the Gregorian Calendar year 2012 – specifically the date December 21, 2012 as the end of a "Great Cycle" of 5,125 years of the Long Count calendar.

The Spiritual Significance of the Mayan Calendars

The 9 Mayan Underworld Cycles depict the evolution of the world since the beginning of (cosmic) time: Each cycle has 7 days and 6 nights that are somewhat similar to the days and night on our planet with more light pouring in during the days than the nights (where we integrate what has happened during the day when we sleep) - repeating themselves at a higher level of creation in waves where each day represents a certain part or path in evolution at that particular underworld level reaching for ever higher evolutionary and shorter time span levels as time and evolution accelerates.

These universal cycles of creation are perhaps parallels to human life:
When a baby is born it comes out in the world and goes through various stages of child development before it becomes a teenager and adult, retires and gets old. The older we get the more we evolve through our earthly experience and potentially spiritually wiser, we may become. The evolution of human life and that of mankind is perhaps a minute, holographic or symbolic "mini-universe" representation of the evolution of universal creation cycles throughout time. The Maya were well aware of this through their recording of cycles and thanks to Mayan researcher and scholar Dr. Carl Calleman we have been able to decipher and recreate them in our time.

The Mayan Underworlds and the Cycles of Creation

The first cycle Cellular cycle started 16.4 billion years ago. Theme for this cycle is action-reaction. A day/night was 1.26 billion years in length.

The second cycle Mammalian cycle started 820 million years ago. Theme: Development of mammals. Each day/night lasted 63.4 million years accordingly.

The third cycle Familial consciousness cycle: Lasted 41 million years. Each day/night lasted 3.2 million years

The forth cycle: Tribal consciousness cycle: Lasted 2 million years. Each day/night lasted 160.000 years.

The fifth cycle: Cultural consciousness cycle: 102 thousand years. Theme: Putting down the tale and picking up the mind as a survival tool. Each day/night lasted 8000 years.

The sixth cycle: National consciousness cycle: 5 thousand years from 3115 BC. Theme: Developing agriculture and law. Each day/night lasted 400 years.

The seventh cycle: Planetary consciousness cycle. 247 years from 1755 AD. Theme: Industrial revolution. Each day/night lasted 20 years.

The eighth cycle: Global/galactic consciousness cycle from 5th Jan 1999. 12 years. Theme: Emergence of global consciousness. Emergence of the world wide web. Each day/night lasted 360 days.

The ninth cycle: Universal consciousness cycle from 10 Feb 2011-21 Dec 2012: Theme: Understanding of human civilization in relation to all of creation, the cosmos and the universe. Big Bang - experiments. Each day/night lasts only approximately 20 days.

Consciousness' Evolution of Each Cycle of Creation

9th Cycle (Universal)	Conscious Co-creation
8th Cycle (Galactic)	Ethics
7th Cycle (Planetary)	Power
6th Cycle (National)	Law & Punishment
5th Cycle (Cultural)	Reasoning
4th Cycle (Tribal)	Similarity/Difference
3rd Cycle (Familial)	Stimulus/Individual Response
2nd Cycle (Mammalian)	Stimulus/Response
1st Cycle (Cellular)	Action/Reaction

Each cycle has a different evolutionary theme and builds on the previous one. The cycles complete themselves exponentially faster and faster similar to the layers in a pyramid (see following page) as time "speeds up" and consciousness' evolution happens faster.

If we relate these universal creation cycles to different aspects of creation consciousness we get a nine step pyramid of the evolution of (human) consciousness through the Mayan Cycles of Creation similar to pyramids built by the Maya themselves.

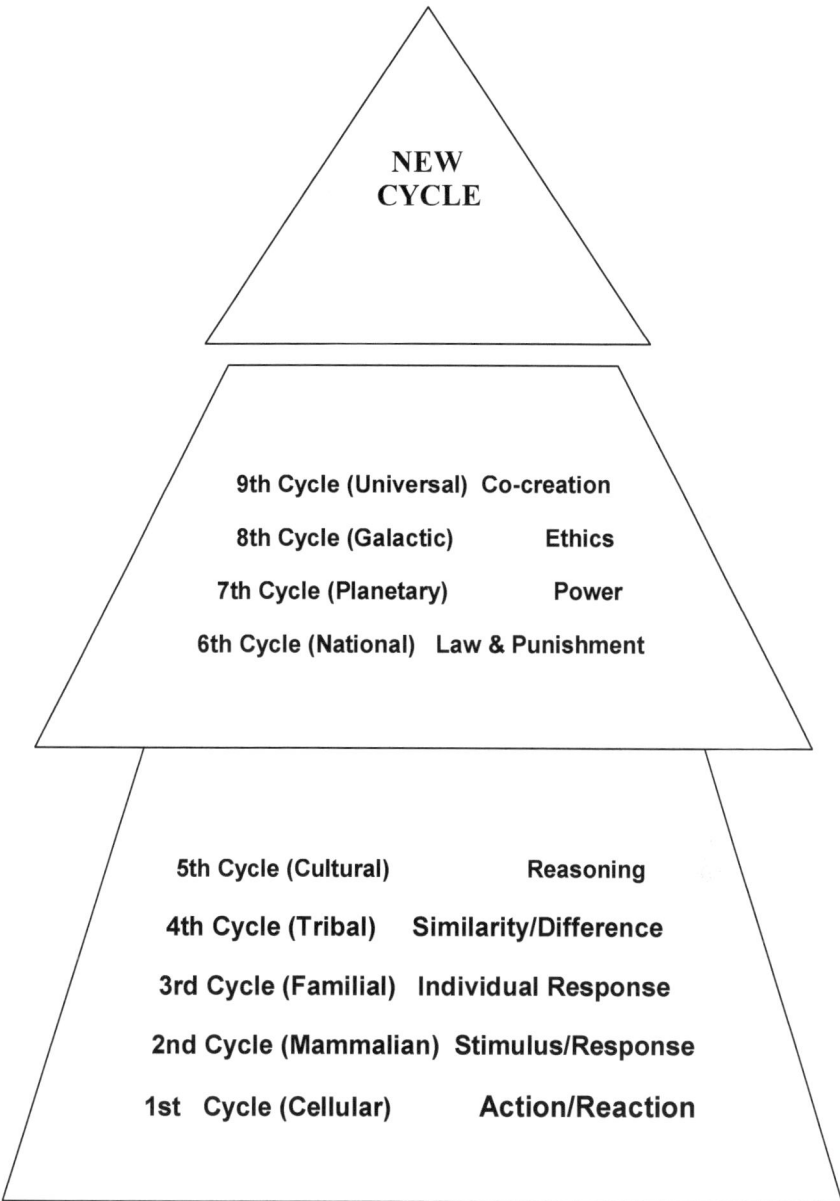

NEW
CYCLE

9th Cycle (Universal) Co-creation

8th Cycle (Galactic)   Ethics

7th Cycle (Planetary)   Power

6th Cycle (National)   Law & Punishment

5th Cycle (Cultural)   Reasoning

4th Cycle (Tribal)   Similarity/Difference

3rd Cycle (Familial) Individual Response

2nd Cycle (Mammalian) Stimulus/Response

1st Cycle (Cellular)   Action/Reaction

What we call the Big Bang which happened approximately 16 billion years ago in linear time is associated with the 1st Mayan Cycle of creation and the Mayan creation myth, the Popol Vuh. During and after the 9th creation cycle we have the potential to make a quantum leap in creation consciousness as we move onto a whole new cycle.

Such cycles end with the destruction of the "old sun cycle" or way of life and the inception of a new world based on the Sun's new placement in the greater cosmos. Many scholars on the subject agree that the Classic Maya pointed to this time, around the year 2012, as the juncture between our present world age and the next cycle. The Mayan Calendar, however, does not go into explicit details about the new one. We therefore have to look to modern day research, science and the surviving descendants of the Maya, present research experiences etc. for further answers.

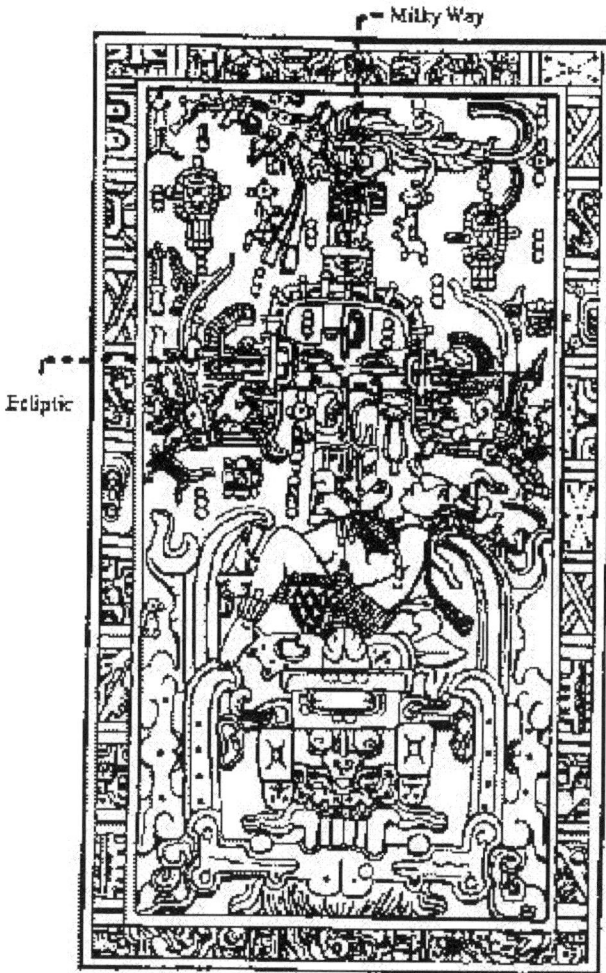

Milky Way

Ecliptic

The Maya Tree of Life showing the intersection of

the Sun's ecliptic with the Milky Way or Galactic Plane.

The Different Meanings about the End of the Calendar

The Mayan calendar in terms of counting ability is so accurate, that it has been able to predict lunar and solar eclipses in relation to the Gregorian modern day calendar. The translation along with the Mayan understanding of time as cyclical, meaning return and cycle back, has caused many of the world's leading astronomers, researchers and scientist on the Mayan Calendar subject to believe that the purpose of the Mayan Long Count is far more significant than the fact that it ends on the Gregorian calendar date of December 21, 2012. Some even say it might not happen in 2012.

The theories that have sprung up about this end date range from the laughable, to the religious, to the scientific. So let us briefly explore some of the views to see what might happen around the magical end date of December 21, 2012:

Prophecy as a Warning
Our population is rising as are the demands on our environment which is depleting our resources. Many believe that we can't carry on the way we are. Could the history of the end of the Mayan civilization then repeat itself? Is humanity or the Earth experiencing some of the same problems the Mayans might have done, or are we reaching a new evolutionary platform? If the Maya were living on an ecological edge in terms of exploiting their environment and having a huge population base, all it would have taken was a sudden, little imbalance in the system for the entire system to fall apart. If we throw in warfare, it is a dangerous mix that could have led to their rapid decline or to the collapse of their civilization. The descendants of the Maya, however, have no such record and question this to be the reason why they vanished.

Magnetic Field or Pole Shift
Another theory that has sprung up suggests that a Magnetic Field or Pole shift could occur around this time because the calendar was based on pole shifts, which have occurred repetitively throughout the Earths history. Could the Mayan understanding of time periods between these shifts have created the Long Count Calendar and the final date be the time for the next pole shift? If so, how much time was supposed to pass between these shifts and how did they know it? Was their record passed down to them from long ago, recounting a time when there was a prior pole shift so that mathematics could be used to predict the next one? Or did they study climate changes by looking at tree rings? We now know that Pole shifts occur with wide intervals lasting up to more than a

hundred thousand years, so this theory is unlikely to explain why the Maya were so preoccupied with time.

Change of Mind or Consciousness

Others, who have studied the Long Count in depth - particularly one leading researcher and scientist, Dr. Calleman - believe the Mayan Calendar predicts a change of consciousness or "enlightenment" in this current time frame suggesting a much more mythical or spiritual approach. This theory coincides better with what we know about December 21, 2012. The Winter Solstice provides us with a view that will not be seen again in our lifetimes. The Sun will conjunct the intersection of the Milky Way in the centre of the ecliptic, giving us a view of the Sacred Tree of Life – as called by the Maya (see previous).

Although 2012 is literally etched in ancient stone, what we do know then suggests that the Maya were offering a warning or a guide to this time – as a time of illumination - rather than a (doomsday) prophecy.

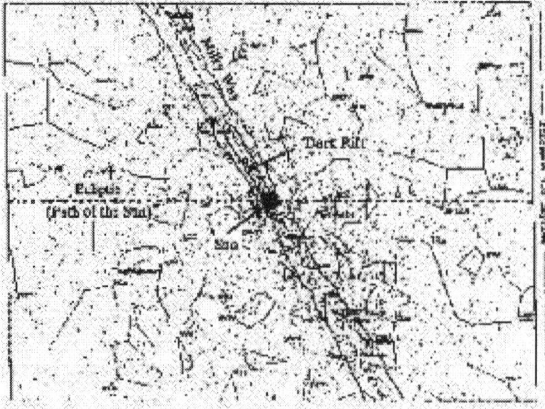

On the winter solstice on December 21, 2012, the sun exactly conjuncts the crossing point of the sun's ecliptic with the galactic plane or equator, while also being closely conjunct the exact centre of the galaxy (top). "The Dark Rift" with the Galactic Centre & equator in the middle (below). Image credit: ESA

The Hunab Ku

The Maya discovered that the entire solar system moved. That even our universe has moving cycles. These are repetitive periods which begin and end like our day and night. Their discoveries lead to the understanding that our solar system rotates on an ellipse that brings our solar system closer and further from the centre of the galaxy and its centre. In other words, our Sun and all of its planets rotate in cycles in relation to the centre of the galaxy or Hunab-Kú, as it was called by the ancient Maya, which is the central light of the galaxy (see above). It takes 25,625 years for our solar system to make one cycle on this ellipse. The cycle is divided into two halves similar to our day and night. The half closest to the central light, is our solar system's 'day' and the half furthest away is its 'night'. Each day and each night lasts 12,800 years.

The Maya discovered that every grand cycle has its minor cycles that carry the same characteristics. One galactic day of 25,625 years is divided into five cycles of 5,125 years. The first cycle is the galactic morning when our solar system is coming out of the darkness to enter the light. The second cycle is the mid-day when our solar system is closest to the central light. The third cycle is the afternoon when our solar system begins to come out of the light. The fourth cycle is the late-night when our solar system has entered its furthest cycle from the central light. And the fifth and last cycle is night before dawn when our solar system is in its last cycle of darkness before starting again. This is the cycle we are currently coming out of.

The Mayan god of all gods, Hunab Ku, is said to exist in the centre of the galaxy, radiating its intent out to life through each local star. Including our sun, as far as the Maya could see. It is in this galactic core that the motion of the galaxy is initiated and its superstructure distributed out to its components. Galactic time and its dynamic web of gravitational effects entrain solar and planetary time which, in turn, entrains cellular time through the circadian rhythms of nature. The Maya were the emissaries of Hunab Ku, tasked with tracking and calculating the movements of time as they relate to our planet and what they might have felt was a kind of "harmonic relationship" with the galactic core as they studied it.

It is interesting that the glyph of the Hunab Ku is linked to the concept of a black hole (or dark rift), which we now know lies at the centre of the Milky Way. Like a yin-yang symbol, the Hunab Ku is a dynamic interplay of black and white. Yin or yang - light or dark. Science tells us a black hole is an ultra-dense gravitational object so powerful that light itself cannot escape its surface. While the galactic core keeps the dark firmament moving and hung with bright stars, their light is always being drawn into the darkness. The glyph of Hunab Ku is said to be a two-way street, allowing access to the core of any galaxy. Warping and bending with gravity, at the point of dissolution time ceases, space collapses, and infinity is revealed.

Mayan cosmology is intrinsically bound to time and the Tzolk'in is their codification of a fractal, cyclical conception of motion as it radiates outward from the galactic centre, through our star to the Earth. The product of a 365 day solar/lunar calendar and a 260 day sacred calendar, the Tzolk'in presents a great cycle of 5,200 tun of

360 days each. This is roughly 5128 years for the current cycle, beginning at about 3113 B.C. Within this cycle of 5200 tun are 13 baktuns, or eras, of about 144,000 days each. The present baktun is the 13th and last, set to end on that fateful day in December of 2012.

Essential to an understanding of Mayan time then, is its cyclical nature. Like the rings in a spiral or galaxy, the events of a day within one baktun will be reflected in the same day of another baktun etc.

Both timepieces, at the very least, suggest that time is nowhere near as linear as we would like to think. In a way it seems far more elegant (or cosmically romantic perhaps) to imagine a resonant fractal harmonic superstructure of time set and stamped by the rotation of our galaxy and distributed to Earth by our Sun star. What happens in 2012 when the entire cycle resets itself and a new cycle begins? Only time will tell...

One thing is certain, the Maya established the first and most complete understanding of universal time in relation to our planet and solar system. While western science put a man on the moon in 1969, the Maya were thinking galactic about time way before the birth of Christ! Within the complex codices of the Tzolk'in appears to rest a template for eternity. Where time, like the flow of a river, appears linear on the surface, only to reveal complex dynamics, eddies, vortices and currents - whirling beneath.

Chichen Itza temple of kukulcan serpent with 9 steps representing 9 cycles of evolution of time. Image credit: Bjørn Christian Tørrisen.

Like standing on the top of the temple of Kukulkan serpent with modern day tourists to the left and ancient Mayan priests to the imaginary right, all of time is ever present, compartmentalized and portioned out only by the fragmented self struggling to make sense of the vast data of hyperspace. Like a tessera, the structures within time remain, only rotating in a higher dimension, archaic but entirely new. Within the Hunab Ku, the black hole, the vortex, as above so below, the singularity of space time - galactic or personal - collapses temporality into the timeless moment of eternity. The serpent gains wings, flying feathered towards the sun with a blackened silhouette against a fiery white star.

According to Jose Arguelles, author of the Mayan Factor, like the I-Ching, the system of Mayan science is one of holonomic resonance. It is as much of the future as it is of the past. Indeed, from the perspective of Mayan science the terms future and past are of little value as gauges of superiority of progress. For the Maya, if time exists at all, it is a circuit from whose common source future and past flow equally, always meeting and being united in the present moment. This is important, if we are to remember our true relationship with time.

CHAPTER 2

ENTERING OUR GALACTIC DAY

Many are now aware the Mayan calendar but not many people truly understand what it means to us that the Mayan Calendar ends on December 21, 2012. What is going to come out of it – when we enter our Galactic Day in 2012? Well, the truth is we have been on this ending path of the Calendar journey for a number of years, already.

Consciousness always has an orientation. In current civilizations, it is in a time and place. Time is an agreement between people - it is an abstract idea, and the agreement is within a calendar. Place is also an agreement between people - every place has its own consciousness. Therefore, the centre point of time and place is where our consciousness is. Our consciousness is our viewpoint and the viewpoint is where you believe you find yourself – right now – is that not right?

If we are to understand it further or more deeply, however, we may need to take both a rational-scientific and non-linear or spiritual approach. The Maya had a very precise understanding of our solar system's cycles and believed that these astronomical cycles coincided with the spiritual evolution of human consciousness. On the other hand most of modern day science agrees that we must use empirical methods to consciously understand reality, but will they help us find explanations or scenarios about what will happen to or within us around or after the end date?

Later chapters of this book will walk through some of the main details of this aspect of the 2012 transition: How the transition takes place not only from an astronomical perspective, but what it means to us spiritually on a more personal and global level, but for now let us move deeper into the explanation of these cycles.

The Mayans prophesied that from 1999 we would have 13 years to realize the changes in our conscious attitude and possibly stray from a path of self-destruction and instead move onto a path that opens our consciousness to integrate us with all that exists.

The Mayans knew that our Sun, or Kinich-Ahau, every so often synchronized with the enormous central galaxy and from this central galaxy received a spark of light which caused the Sun to shine more intensely producing what our modern day scientists call solar flares as well as changes in the Sun's magnetic field. We now know, the Maya believed this happened every 5,125 years and this might cause a displacement in the Earth's rotation because of great stirrings in the Earth's crust, but would catastrophes be produced as many are saying?

The Mayans believed that universal processes, like the 'breathing' in and out of the galaxy, are cycles that cannot be changed. What potentially changes is the consciousness (in greater terms) that passes through it in a process toward more understanding or perfection. Based on their observations, they predicted that from the initial date of the start of their civilization, 4 Ahau, 8 Cumku which is 3113 B.C., after one cycle being completed 5,125 years in their future, on December 21st, 2012, the Sun, having received a powerful ray of synchronizing light beams from the centre of the galaxy around this time, could change its polarity to produce a great cosmic event that would then propel human kind to cross into a new era. It is after this, that the Maya say their calendar ends and why we will be the ones to go through the door that was left by them, potentially transforming our civilization into a whole new age.

Only from our individual efforts could we avoid a path of catastrophe or cataclysm that our planet or civilization might suffer before the start of a new era or cycle of the Sun. The Mayan civilization, and there were four other great civilizations before them, were destroyed (by great natural disasters). They believed that each cycle was just one stage in the collective consciousness of humanity. Like a death before a rebirth.

In the last "cataclysm" of the Mayans, their cities were suddenly abandoned at the height of their civilization. They believed that having known the end of their cycle, mankind would prepare for what is to come in the future. If we are to preserve ourselves; the human race, as the dominant species, the coming changes might permit us to make a quantum leap forward in evolution to create a new or more evolved civilization that could manifest greater harmony, love and compassion for the benefit of all humankind.

Their first prophecy talks about 'The Time of No-Time' a period of 20 years, which they call a Katún. The last 20 years of the Sun's cycle of 5,125 years. This cycle is from 1992 - 2012. They predicted that during these times, solar winds could become more intense and could be seen on the Sun. This has already happened and would be a time of great realization and great change for mankind. It would be our own lack of preservation and contamination of the planet that would contribute to these changes. Accordingly, these changes would happen so that mankind would comprehend how the universe works so we could advance to superior levels and liberate ourselves from further suffering.

Seven years after the start of Katún, which is 1999, we would enter a time of darkness which would force us to confront our own conducts. The Mayans say that this is the time when mankind will enter The Sacred Hall of Mirrors where we would look at ourselves and analyze ourselves and behaviours with others, with nature and with the planet on which we live.

It is a time in which all of humanity, by individual random or deliberate conscious decisions, could decide to change and eliminate fear and lack of respect from our relationships. The Mayans prophesied that the start of this period would be marked by a solar eclipse on August 11, 1999, known to them as 13 Ahau, 8 Cauac. This would coincide with an unprecedented planetary alignment, the Grand Cross alignment. This would be the last 13 years of the Katún period with an opportunity for our civilization to realize the changes which are coming at the moment of our spiritual rebirth or renewal.

The Maya predicted that potentially along with the eclipse, the forces of nature would act like a catalyst of changes so accelerated and with such magnitude that mankind would be powerless against them. Also, that our technologies in which we rely on so much would begin to fail us if we refuse to learn from our civilization and the ways that we are organized as a society. Our internal, spiritual development would require a better way to interact with more respect and compassion.

The Mayan prophecy tells us that in 1999, our solar system began to leave the end of the cycle which started in 3113 B.C. and that we find ourselves in the morning of our galactic day, exiting darkness and on the verge of being in plain day of our central galaxy in 2012.

They say that at the end of this cycle the central sun or light of the galaxy emits a ray of light so intense and so brilliant that we cannot escape it – because it illuminates our "universe". It is from this burst of light that all of the Suns and planets sync. The Mayans compare this burst to the pulse of the universe, beating once every 5,125 years. It is these pulses that mark the end of one cycle and the beginning of the next. To the Maya everything was numbers and if we use the Mayan notion of 13 katun years that gives us from 1999-2012 where the exact middle point of the alignment takes place, with another important 13 years for the future period from 2012-2025 when the calendar potentially reboots itself.

As we come back to what they call 'The Time of No Time', this is a highly evolutionary period, short but intense, inside the grand cycles of time where great changes take place to thrust us into a new age of spiritual evolution as individuals and as mankind.

What Does It Mean to Reach the End of A Cosmic Cycle?

Cosmic cycles on the order of thousands of years are converging now, in our lifetime. The 2,000-year Age of Pisces is flowing into the Age of Aquarius. An "astrological great year" is ending as we approach an astrological "new great year". The 5,125-year Mayan calendar ends in 2012 as our Sun crosses the galactic equator at the December Solstice ushering in an era of change and new opportunity on Earth.

Modern astrologers are well aware that it has special implications that we are on the cusp of two astrological cycles as we are moving from the Piscean to the Aquarian Age. Even in 2012 we are still slowly moving into early Aquarius, so the new astrological cycle has not fully emerged from astrology's point of view. Although we have been on the cusp for several decades since the late sixties, it will not be until we are well into to the second millennium that we would really start to experience the Aquarian Age full scale. In this way we are neither in the Piscean nor the Aquarian Age, but find ourselves in a time of transition furthered by many extraordinary planetary alignments. Whether we believe in astrology – coming from a scientific point of view or not – it has been experienced by people for hundreds of years as a valid, useful tool or indicator of a connection between the cosmos and our daily human existence here on Earth.

The idea of an ongoing 2012 shift – spanning several decades, comes from many different past and contemporary sources – although the Mayan prophecies are perhaps the most precise and encompassing. The fact that so many cycles are converging now, suggests that we are not just on the threshold of a new world age, but poised for a quantum evolutionary leap in the next few decades as these cycles converge or overlap. This is a vast, uncharted terrain. In a way we are stepping off a precipice and the future is yet unknown. However, as change is the essence of evolution, and evolution is speeding up around this time, there is and can be no stopping or slowing down of the process. We came here somehow to experience the journey or ride at this very special point in cosmic time. The more conscious we become of the non-linear, synchronistic nature of existence, the more rapidly we may indeed evolve beyond the linear mind of the old time paradigm.

What Is the New Cycle of the Sun?

What exactly does it mean to enter a new Sun cycle - you may ask? A cyclical understanding means that endings are rarely barely endings. If we consider that the universe and life here on Earth are connected it could mean a death-rebirth-new life process - just as in the natural web of life. Even Einstein proposed or justified some of these ideas through the theory of relativity – where energy and matter – essentials to all life - are somewhat interchangeable, meaning the visible and invisible aspects of our reality are two sides of the same coin or can be replaced. The universe is certainly a vast, complex place where many things are in a constant state of change, flux or motion - resting somewhere between predictability and unpredictability - as in chaos theory perhaps.

More recent models of the universe – like that of Stephen Hawkins' - suggest that huge clusters of suns or stars can be sucked into a black hole and disappear as the universe turns its head or tail upside down in a reversed or opposite kind of equation to recreate itself on the other side.

On a more local level we can relate this to our Sun's galactic alignment. Although the Sun is located on a distant spiral arm far from the Galactic Centre – and we may not be able to understand the Galactic Centre because it is far away, as the Sun aligns with it, one cosmic time cycle may end and another begin as the Sun passes through the Galactic Centre's enormous gravitational forces of light and energy. A galactic alignment is a very powerful location for absorbing source light or energy from the Galactic Centre at its waves point outwards through space and time to affect other parts of the galaxy over very long distances.

Not surprisingly, we have been able to retract more solar flares, wind or particles spewed out from the Sun before they hit the Earth's magnetosphere around the poles and are turned into magnetic field disturbances or northern lights.

Similarly, our solar system might also be affected like the Earth by the "cosmic atmosphere". Perhaps this is one of the most overlooked scientific surprises or discoveries of the new millennium in understanding why we are seeing planetary temperature rises over the 2012-Sun-GC alignment not just here on Earth.

The Maya associated this 2012 phenomenon with what they called the Xibulba or the black road. Although they did not posses the scientific knowledge or instruments of today to detect it, they were still able to accurately see and measure its passing in the sky during the precession of the equinoxes where the Sun circles around the backdrop of the Milky Way's centre.

If we take the Mayan view of Earth in relation to Creation into account, this could mean the end of a vast cosmic cycle for the Earth in the cosmos. Indeed, our whole planet is an integrated part of the Solar System, so the whole solar system might be going through a similar type of timely transformation.

As the Sun is affected by the Galactic Centre, we may be throwing off whatever cosmic or energetic debris we have carried with us from the old time cycle as we see the planet also "clearing" itself – through earthquakes, hurricanes, floods etc. as depicted in the Dresden Codex.

Just as important the years leading up to and around 2012 will be – the years and decades after may become equally or perhaps even more important as a whole new cycle for the shaping of our planet and its future begins. What happens during this time could have far reaching implications or consequences for us all.

Although our Sun and planet has gone through shifts of cycles before – the difference this time around is that this is a much bigger one because we now live and breathe in a huge modern civilization the world has never seen before. The last time the Sun aligned with the galactic centre predates historic records, so we have nothing to compare this point in our human history with. Perhaps the latest records of such a shift were lost with the destruction of Atlantis – a possible, prehistoric, advanced human civilisation. Many clues point in this direction although it is not possible to go into further details about it here.

What this time frame means is of course yet to be experienced as we measure it with our clocks. It will of course not cease time as our clocks do not stop, but time by physical measurement in a linear fashion may change or bend and with it perhaps our perception. We all now what it feels like when a day passes by like an hour or a week drags along slowly like a month. Time is not always what it seems.

As the intense cosmic energies are destined to inundate our planet, the influx of energy may not only cause disruptive changes in our space-time weather perception. We may also become more susceptible to what goes on in the outer universe – through our timeless consciousness or universe within. Are we ready to move beyond outdates ideas, old thoughts or concepts through time as we have come to know it?

Evidence of 2012 Date on Ancient Mayan Monuments

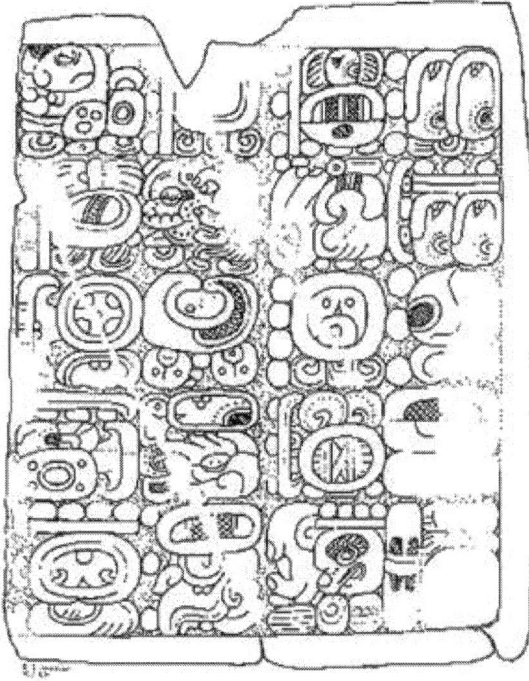

Talking about 2012, there is only one ancient stone monument with Mayan inscriptions that depicts the end of the current era in 2012: The Tortuguero Monument #6, is erected around 670 AD in Tabasco, Mexico. By decoding this inscription, archaeology has been able to determine that the end of the 13 Baktun Cycle Long Count Calendar Cycle is December 21, 2012; 4 Ahau. This ancient T-shaped monument links our modern world to a larger, mythical story of unfolding. The monument is reported as originally being covered with inscriptions. Shown here is the right-hand "wing." Sadly, the left "wing" of the monument is missing, the central section is partly effaced, and the part where the prophecy is recorded has a big crack through it, making full translation virtually impossible! This is why we have to rely on accounts from Mayan elders, scientific and other sources to understand the time we are in.

Bringing What We Know About 2012 Together

From the Mayan prophecies to the indigenous prophecies of North America, our ancient traditions suggest that something really big will happen during our time in the Earth's history. Although nothing in any of these prophecies tells us that the world itself will end, what they do say is that the world as we have come to know it will enter a time of change in years around 2012. How we respond to that change will define how we experience it and our lives in the next age or cycle of our existence.

What sets the Mayan prophecy apart from the general predictions of other cultures is that it has an expiration date that occurs in our time which is really significant. The last cycle of the intangible Mayan calendar corresponds to a series of tangible events, some of which are happening, today. Here are some of what we know for certain:

- The winter solstice of December 21, 2012 is an alignment so rare, that we have been "preparing" for it for more than 5,000 years, and it could be another approximately 26,000 years before the same opportunity cycles around again.

- A cycle of solar flares or storms are predicted to peak in 2012-2013, with an intensity that could be 30–50 percent greater than previously. NASA has warned it could knock out our telecommunication systems.

- Many scientists agree that the Earth's magnetic field is weakening. The presence of a weakened magnetosphere could prime us to accept new universal ideas, bring changes or experiences depending on how susceptible we are to the power of this "synchronization beam" relating to the cosmic changes.

- While solar cycles, magnetic changes or reversals are very real and have definitely happened in the distant past, a galactic alignment like the one surrounding 2012 has never happened with 7 billion people inhabiting the planet who depend on technology of power grids, internet communications, computers and global positioning satellites.

- Climatic and Earth changes have been happening at an increasing rate. Although many scientists indisputably say that climate changes are man made, other Earth changes such as more quakes and tsunamis might not be. Also, scientists have discovered recent climate changes and temperatures increasing on other planets in our solar system as well.

- We know that the Mayan astronomers have accurately depicted an important Venus Passage that will occur just after Dec 2012. The ancient Maya often associated this with war. It is a warning.

- Validation from quantum physics and experiments suggests that the way we perceive our world through our thoughts and emotions – greatly influences our physical reality.

With these facts in mind, the December 21 solstice of 2012 seems to be of great significance and as the research and understanding of the Mayan Calendar suggests, it opens a rare cosmic window of opportunity in time with a change point or threshold we must pass through where things click into place which allow us a new way to see ourselves in the universe in relation to our state of being and human consciousness. Just as uncertain the future of our planet may seem, we simply do not know what passing through such a monumental experience could mean to our human life and bodies.

Although a time like this is surely not "business as usual," the fact that people lived to record some of what happened during past cycles tells us that such events are manageable and survivable.

Our rendezvous with 2012 suggests that if we live life focused upon all of the bad things that may happen, we will miss the joyous experiences that may keep those bad things from actually happening! The stage is set, the opportunity to experience it and choose what we will do about it all is ours.

Will we acknowledge or ignore the importance of what is happening "out there" in relation to us "down here"?

Linking the Earth and the Cosmos

The Maya came through direct experience to believe in a strong relation between the Earth and the cosmos because their exploration of life and studies of the universe told them so. As they studied the movements of the celestial bodies, they discovered that it was more than simple distant or dislocated movements of stars and planets. They became aware that planetary passages could have an impact on climatic patterns and weather conditions which in term affected their crops, culture and civilisation which reached a climax between 300 B.C. and 900 A.D. That planets and stars affect human life here on Earth is nothing new - it is something both ancient and modern astrologers adhere to. Environmental scientists also know that slight changes in the Earth's atmosphere can disrupt weather patterns that affect millions of people's livelihood around the world - through droughts or floods - as those we are seeing now.

Astrophysicists and scientists are also aware that solar flares have a direct impact on the Earth's magnetosphere. Modern science has even measured the physical impact of the outer planets. Jupiter's gravity for instance, holds the asteroid belt outside Mars in place, so that we do not get bombarded by giant rocks from out of space into the Earth's atmosphere.

It seems somewhat paradoxical then that we have all this scientific and other knowledge yet our scientists including NASA's say that the heavenly bodies around 2012 – will have no significant impact. What do they know? Of course we must listen to science, but if we follow in the footsteps of our ancestors through archaeology, it does seem like a great understatement that 2012 will have no detectable impact. Whether we are coming from a rational-scientific or more spiritual point of view does not matter. We should continue to keep our minds open to the fact that life here on Earth is somehow related to what goes on in the inner and outer rims of the solar system and in turn the greater cosmos. Even the most sceptical astronomer or scientists will agree that anything, but further scientific investigation can be fatal or stupid. After all the Earth's atmosphere is only a thin penetrable layer – similar to the outer shell of an egg and we all know that without the Sun, life here on Earth would most certainly cease to exist.

The split between science and metaphysics or spirituality is perhaps one of the biggest modern day quandaries that we are faced with reconciling if we are to understand the Mayan Calendar and its concepts fully and may be bridged in relation to the Great Shift of the Ages. What is greatly needed is that both modern day science and metaphysics put their cards on the table if we are to understand the time around 2012.

The idea of a direct link between human life on Earth and the cosmos may contradict what most of modern day science have to say, but if we sit around and wait for science and rationality to catch up with our spirituality it might become too late........?

The extraordinary planetary alignments surrounding 2012 have caused modern astrologers to search for a deeper understanding or meaning. Many of them consider particularly the grand T-square between Saturn (old structures and ideas), Pluto (the planet of spiritual transformation and change) and Uranus (unpredictable change, new ideas or "awakenings") an epitome or a sure sign of change and challenge surrounding the battle between old and new surrounding the time where the Mayan Calendar officially ends.

Some astrologers believe this could make the social and societal changes of the 1960'ies look like a walk in the park - equivalent to the difference between spiritual evolution and revolution. Forty years later, one decade into the 21st century, our world still upholds beliefs that shaped the last millennium. Now perhaps more than ever, we need to see life differently.

We are in the midst of a worldwide social & cultural reality check and revolution. The square of Uranus to Pluto will not permit us to live as we have been living before, neither individually or collectively as people of Earth.

The square from Pluto will create greater urgency and pressure for conflict or war in order to achieve the goal of change and transformation into a new way (Uranus) of living in the structures of societies (Saturn) in the world. This fifth phase of the Saturn/Uranus cycle is actually the harbinger of two Pluto cycles: the square from Saturn to Pluto (visiting three times between November 2009 and August 2010) and the square from Uranus to Pluto which will come within winking distance in July 11 but not perfect until June 2012, continuing through 2015. What happens during the Saturn/Uranus opposition will set the stage for the next eight years or so.

The T-Square is also showing how we need to change our (economic) systems, and suffer the inevitable growth pains of this change. Outdated or old ideas will probably not survive "the tomorrow".

Knowing this information is only useful if we do not use it to scare ourselves to death. It is important to view these cycles within a framework of faith and understanding so that we can use it to our advantage and not become overwhelmed by it. The keynote of this entire period could eventually be the transformation of our existing scientific, religious and political (belief) systems. What do you believe?

It is a time to examine our own lives to see where our life is not authentic or where we are not being true to ourselves and where we need to go with changes (Uranus) that must be made to the structure of our lives (Saturn), as a result. it is the perfect time to cultivate faith in the natural flow of life or the universe and to increase our understanding that we are spiritual beings having a human experience - not the other way around. It is a wonderful time to learn to let go when doors to old desires are closing, and to remain strong in our intention to manifest a life of beauty and adventure that resonates more with what we feel inside. This, after all, is the ultimate challenge of any planetary and evolutionary cycle in astrology.

Combining research on astrology, astronomy and the Mayan Calendar system has given us many clues about 2012, but the end of the Long Count Mayan Calendar has still left modern scientist, researchers and spiritually inclined seekers alike baffled while trying to understand what the link between the galaxy, our solar system and time is really about.

Our Relationship with Time

Unlike the Mayan Calendar, our Gregorian calendar is a linear approach to time and how it unfolds – so what if we need to look at how we are currently using time in our daily lives, or how we could use it differently; in relation to the past, present or future perhaps?

The days and years in the Gregorian calendar progress with no particular cosmic or spiritual evolution in mind. This is quite strange from a Mayan perspective. It does not assist us in realising our potential through a timely interconnectedness with the universe. Instead, we have chosen to be bound to the wheels of earthly time through a 365 days-a-year limited calendar system which goes on year in and year out with no particular purpose or reason why we are living in mind, other than simply registering the days, weeks and months – leaving us no clues as to life's purpose or the greater plan of our lives in relation to consciousness or the cosmos.

In the Gregorian calendar human civilisation simply goes on along a linear trajectory - whereas in the Mayan Calendar system time becomes our friend and ally as we travel through life around the sun in our solar system. The galactic movement of our solar system becomes an integrated part of our timely (spiritual) evolution. Do you see?

Linear time as a human creation has human common denominators. These are the digits from 1 to 12 and all the space in-between- the digits of 60 seconds, 60 minutes, and 24 hours. All of these bring you into alignment with the limitations of linear time, but how does this relate to your average day? Well, you get up in the morning and the first thing you do is what? You look at what time it is. You look at the numbers on the clock. At that moment of reference you decide how much time you have left to make coffee, to eat breakfast, to shower, to get dressed – to get to work and so on.... like the wheels of a thread mill ....life goes on in the same old circles. Rarely do we wonder, shall we stop here or there? Instead we speed along because there is not enough time. Almost dangerously, you may wonder - will I get there? In this time sequence, you are cutting yourself short by trying to fit yourself into 60 seconds that make up 60 minutes that end up with 24 hours a day. One day after another these numbers on our clocks become our lords and masters...

In this way linear/calendar time locks us into a little black box that keeps us scurrying about as a rodent in a maze where there is no solution and no cheese - right? What if the essence of time was different? What if you were commander-in-chief of time – where time obeyed you instead of you obeying time?

Do you see how very freeing that thought is? Do you realize how much life force and energy we can waste trying to fit into someone else's time pattern – obeying the dictators of time? What if in reality, time were yours for the bidding and it does not matter what the clocks tell you? If linear time is not composed of matter, but is an illusion, we can understand time in a different way and begin to sculpt life more according to our timely needs or wishes: Create an outcome or game plan that is better aligned with our soul's purpose. By giving time to our inner self by spending time wisely – each day - with people that we care about or our children, time can become our treasure.

Most people just spend time like money. They do not realize it, but suddenly it is all gone. Do you see that when you spend time in this way, you spend prayers, energy and your essential light or life force? You spend it – everyday, but what if your days are not spent thrifty at all? How would you really like to spend your dinners? The essence of time, life, energy and consciousness spent – go hand in hand more than we tend to think. The thought, that money is time, and time is money – so you should be paid appropriately for the time that you spend, casts its long shadow over our lives and earthly events. We have become slaves to a life rhythm that we did not necessarily agree to in the first place.

What if abundance of time was your birthright, earning a living and receiving was generously given? Think about it for a moment.....

If you see your days locked into an approximate 8-hour workday all the time, the way you spend your time – spends you – you make a living, but what if it is a picnic lunch instead of a grand buffet? By stretching the perimeters or parameters of the ways we spend time, we can extend love, energy and money by shifting our perception of time into a format of having enough time, energy, and money in our personal spaces through the ways we spend time itself. Our space-time continuum can be short and sour or it can be elongated and sweet. We can stretch every aspect of our life in accordance to our wishes if we take heed and note of what is really necessary and most important or needed during the particular moments of time.

The Maya so accurately described this link between the universe, solar, planetary and our human evolution through cycles all the way from their birth and the beginning of creation to modern day human civilisation that it blows your mind. In the Mayan Calendar system life becomes a mystery - an ongoing part of a greater evolutionary experience within creation and the advance of human consciousness through time.

The Mayan Calendar's ancient cyclical understanding of time is in alignment also with the developing modern science's understanding of the universe as a web of energy flowing through all of creation. Sort of like a whirlpool or field of light, dark matter or energy which surrounds us all and that we are all drawing life force from. When we know that the Sun is the centre of this field in our solar system - how might the 2012 galactic alignment affect us? May we realise our true relationship with time? Will it shake up our inner relationship with space and time?

If the cosmos works as a vast interrelated web or sea of energy, as modern science has proposed, where all things are somehow interrelated, how can we take this grand cosmic alignment – for a piece of pie in the sky for future times? Why do we take our day to day assumptions about linear time reality for granted? If we are part of an ultra greater or multidimensional reality – that encompasses far more than we can presently imagine - what does it mean to our consciousness or comprehension?

Would that enlighten our consciousness in relation to our understanding of life here on Earth? What if we were connected to other life forms on outer planets circulating distant stars?

It appears our so-called modern day civilization is still slacking behind the ancient Mayas spiritual or cosmic understanding.....with no multidimensional understanding or universal approach to human life here on Earth.

What of our so-called modern day human civilisation in its quest for scientific answers and down to Earth physical solutions have totally ignored is that the linear rational mind and its perception of time have rendered our soul and its spiritual evolution useless. If our souls in relation to God, The Great Spirit, The Cosmos, Universe or whatever we want to call it is (with)in our hearts and related to our minds as well, we are truly intertwined through a vast complexity of universal life with more between Heaven and Earth than is apparent to the human eye.

If we become conscious of how our physical/earthly lives relate to the great universal organism, we can learn to respect and be grateful for one another and our planet and its life forms once again and possibly delineate a new pathway which could take us directly into a more positive growth pattern that reshapes our cosmic future. The Sun and the Galaxy together with our planet are awaiting our decision(s) as we face a zero point.

Spiralling Through the Zero Point of 2012

Part of entering the zero point around 2012, is that it appears to be a moment in time where linear time could stop or everything start appearing as if it is happening all at once as we become submitted to a quantum wave or phase shift of photon energy from the Great Central Sun of the Galactic Centre. It depends on our personal timely perspective. This photon energy is unlike anything you have ever seen, heard, or been a part of and will permeate anything that comes into contact with its path during The Shift, it will permeate our Sun and the Earth during 2012 in invisible ways. The outer edge or belt of the photon energy has already reached Earth's atmosphere for several years and is affecting everything residing on her. This means serious business.

The Photon Belt - A New Energy for the Earth & Humanity

Because it is entering our part of the universe in unprecedented amounts on a new wave of energy - a future cycle or era will be born in its wake. Photon energy has the potential of aiding the Earth in healing some of the runaway environmental and other savagery problems by assisting us as humanity to move into a new or higher energy frequency spectrum. This frequency has powers to heal, align and lift human things to a higher ground because it affects the human body's molecular structure to realign with more source-light. Photon energy is source light energy that spirals and moves through every galaxy at one point or another. In a sense, the Great Shift is the evolution of Mother Earth and her inhabitants into the realm of photon energy – which potentially affects the next 2,000 years of our evolution.

It is such a powerful light energy or force field that it has the potential to replace our dependence even on fusil fuels for electricity in the new millennium. It is a free energy source that nobody can monopolize and that can be harnessed to meet our future power needs...

Since photon energy vibrates at a very high frequency, it confers the power to bring energy to a system that can no longer operate on its current energy - including our human consciousnesses. Therefore, it is essential to maintain clarity and purity of thought. Those in the process of change and transformation will learn to adapt to photon energy through meditation, being in the "now" and staying heart centred, but is it a spiritual healing energy that works for the highest good of us all?

The actual rate of spiritual "awakening" of the Earth's population is directly influenced by the increased level of photon energy from the belt of the Galactic Alignment around 2012. Humanity's awakening to the spiritual nature of Earth can and will be influenced by this time factor. It is sending us a message to pay attention to the Earth as she is calling for a realignment because of the incoming photon energy.

A Quantum Wave Explosion?

Photon energy comes in waves. The ebb and flow of a wave is normal and natural. When these waves hit us, they are so big that they engulf the Earth plane almost all at once. It doesn't take two or three days to span the Earth. The waves just sweep in and sweep out, quickly. They come in very quickly and leave very quickly particular the closer we are to or around 2012. A wave looks somewhat like a spiral. It does not really have a beginning or end. Only because it sweeps past Earth does it appear to have a beginning and end. The thickness of a wave is important. A thick wave measures at a higher number because there is more photon energy permeating the Earth plane at that time. A thin wave might measure a quarter mile wide, and a thick wave can measure 1.5 to 2 miles wide.

Photon energy amplifies what it "touches" as it literally floats right "through" us/you. When it is in, a person can feel great as if cosmic love has descended on him or her, but it can also amplify our fears if we let it, or make us feel that we are oscillating between feeling exhilarated, confused and depressed as it leaves. The key to working effectively with photon energy is to stay in Truth-Trust-

Passion and not become dependent on it. It is about progress in our spiritual growth and life while we are assimilating the Great Shift.

The Increased Cosmic Light Pulse

As we pass through the Mayan Calendar end date over several years, many things will be affected by the cosmic light pulse emanating from the centre of our galaxy reaching a maximum magnitude on 21 Dec 2012 when reaching into an accelerated time zero point. A zero point in time, according to Terence Mckenna, who first described it, is somewhat similar to a null-zone where the perception of time may come to an abrupt halt, stop or end before it can/will move forward yet again. Terrence McKenna's theory says that time is split between 'novel' and 'mundane' time events: Time gets more 'novel' as it proceeds until everything becomes novel.

A complex computer analysis of the I-Ching revealed a hierarchy of 26 time waves, (each one 64 times larger than the one below), which govern the unfolding of events in the universe.

Mathematical calculations ended the time wave 21,12,2012 for constant novel time (novel instead of mundane time) and this is the end of the Mayan Long Count Calendar.

According to David Sereda, time waves are also emitted by the brain; in lower state of brain wave activity (theta and epsilon), time passes incredibly quickly and it slows down. Huge amounts of time pass quickly.

We might therefore perceive time and with it change to be happening faster and faster till the old cycle closes. In this way the timely Shift will affect all of us.

Although it is a highly complex question what an increased universal cosmic light pulse means to the human mind and body, consciousness changes can affect us in ways outside our rational minds because these waves work from a distance through space in invisible ways.

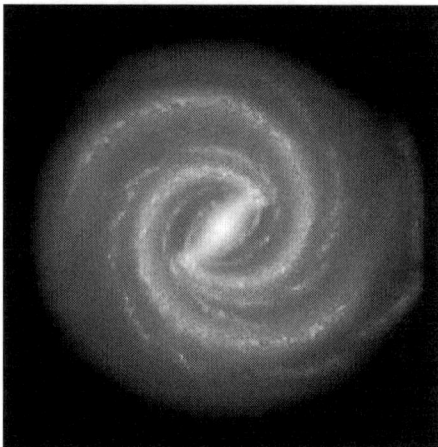

The Messier 74 galaxy (left) with billions of stars forming within its spiral arms (left). The Milky Way (right). Images by NASA, ESA and Hubble Heritage.

The centre of galaxy spirals resemble the galactic alignment as our local Sun aligns itself with the Great Central Sun of the Milky Way surrounding 2012. Even though our Sun is billions of light years from the Galactic Centre and our planet is protected by its magnetosphere, it could affect the physicality of our planet as we pass through this "galactic-core-energy-wave". This in turn could trigger earth quakes or volcanic eruptions as the Earth's body - its tectonic plates need to adjust. Atmospheric weather disruptions could possibly be another reminder that it is more than "climate change". Similar to intense "cosmic storms" in space, we might then experience "storms" here on Earth - including cyclones, snowstorms, wildfires, hurricanes, tsunamis and floods. Whatever or wherever it happens, we have to stay alert to the fact that something unusual might be coming our way during the galactic alignment.

A Cosmic Shift Alert

Scientists have also been witnessing an increase in the Schumann resonance - the energetic pulse of the Earth which used to be at the frequency of (7.4 Hz/sec). Life on the planet measures the passage of time by it and interestingly our brains operate at a similar frequency (in alpha /relaxed) state.

The Schumann resonance which used to be 7.5 cycles per second has increased since the beginning of the 1980s-1990s to a point where it is now above 12 cycles per second. Whether this is due to human or other influences on our planet or both is an open question, but it points in the direction of a direct link between human collective consciousness and the activity of the Earth as we inhabit the Earth's surface. Is the increased Schumann resonance a cosmic shift alert which relates directly to the collective human consciousness and how willing we are to transform, adapt and evolve during the Shift with other species?

If there is resonance or dissonance between the cosmic wave patterns and our consciousness and its brain wave states – what does it mean?

Is our brain experiencing an increased cosmic light pulse? Does it mean we will become more or better aware of the connection between the Earth, our selves and the greater Cosmos or perhaps our Creator? Recent scientific experiments support the theory by pointing us in a direction which indicates that a tremendous amount of information can be stored or contained in a single pulse or wave of light (photon).

The Relation between Galactic Sun Cycles and the Earth

As the galactic alignment of 2012 triggers an increased cosmic "pulse of light" from the Galactic Centre as seen from the Earth through the Sun when it aligns with it, how will this trigger evolutionary changes on our Earth and in human consciousness?

If linear time is collapsing into universal time where things happen simultaneously and we are integrating or processing different aspects of consciousness as we align ourselves into this reality, where are we then - or where will we be after 2012?

Could it be that it will be somewhat similar to moving from slow typewriting to fast computing or from reading an article with a photo to watching a 3D movie - on a consciousness level over a period of time through these years? Our perception or experience of time, space and evolution as we have known it is likely to change similar to moving through the eye of a storm or a tornado with a void in the middle. Faster and faster it goes till we reach the exact centre of the spiral where time or space reaches a stand still before de-accelerating out in a new direction on the other side. The alignment between the Sun and the Galactic Centre is the "centre of gravity". The Great Spirit, light, God or whatever name we prefer or want to use is related to the centre of our galaxy – The Milky Way.

Even if it seems far away– we are as much a part of creation or of our galaxy, as it is a part of us. The only difference being our perception. How far or how close do we perceive it to be - unequal or equal to the physical distance? Does more light mean greater closeness to Creation and with it "Creation Consciousness"? In 2012 it seems we will be closer than we have ever been, but how this will unfold our spiritual evolution is far from set in stone with regard to the future journey of humanity.

The Acceleration and De-acceleration of Time

We now know that a grand cycle of the famous Mayan Long Count Calendar ends close to 2012, but the Sun's alignment with the Galactic Centre stretches over a long period of time. The pi-spiral's centre point may be 2012, when the Mayan Long Count Calendar ends, but what happens as we continue beyond it?

Many researchers on the Mayan Calendar have failed to explore or realise, how important it is that moving towards the end date does not follow a strict linear, but a cyclic pattern towards the Mayan Calendar's end. The changes that will occur from 2009-2012 follow an accelerated logarithmic pattern, the "up shifts" from 2009 alone are greater than the sum of all transitions since 2001. 2010 -2011 further accelerates the process and offer infinitely more through a quantum wave expansion. 2012 then becomes the centre point of the greatest shift in cosmic time – offering the infusion, before it begins to de-accelerate after 2012.

Similar to a very large hourglass of sand, time will be shifting very quickly through it, when the final grains of sand reaches the hole in the middle and then all of a sudden changes half and turns the other way. This means that just when we thought we were "getting it", the sun and universe turns its timely movement upside down and time shifts and begins anew or is different again. What was in time might not be any longer before something new begins. Similar to the hourglass, turned upside down, this is what is happening to our human evolution as we are growing out of an old time cycle and into a new one in life. This may initially be very hard to grasp or understand, at least until we experience it ourselves.

As a climax is reached around 2012 – the Sun appears to then move away from the GC in an inverted logarithmic pattern. Will past, present and future timelines – in consciousness - then be converging? Perhaps this depends on whether we see 2012 as "the end".

If we are going to have more and more time events to deal with – coming from all kinds of directions – the perception of linear time gets compressed – meaning it can be felt or perceived as if there is not enough time - or that time is running very quickly through our days.

Beyond 2012 this perception of time might decrease after the previous acceleration period, but since the Shift is still ongoing changes will be following depending on how we manage to work with it. Like a window blasted open in time these are the years when the world "flips over" into a whole new era.

The Pinnacle of Time Acceleration and Evolution

It is undoubtedly going to be significant and important for us short term or long term because we cannot escape this cosmic evolutionary trajectory we are riding during the shift of cycles.

Ending the Mayan Universal Underworld means we are currently going through the most rapid acceleration in our evolution since the beginning of the Classic Mayan recordings of time, but what is it going to mean to us?

As Mayan Calendar researcher and scholar Carl Calleman has pointed out, the Mayan Calendar is not so much a description or recording of time as it is in fact an unfolding record of evolution as it accelerates within the consciousness of creation. Each proceeding Mayan underworld represents an aspect of evolution or human consciousness with shorter and shorter time periods leading up to the end of the calendar – and the final universal underworld.

Although not all scientists on the subject agree with Dr. Carl Calleman, there is no question about it having certain implications. Just as we have evolved through the prior Mayan Underworlds, a potential, ultimate pinnacle of our human evolution in consciousness through the final days and nights of the Universal Underworld is possible.

According to Carl Calleman, on November 3rd 2010 we entered the Seventh Day of the Mayan Calendar. This New Day period will activate the dominance of the feminine principle (through spirit's higher intelligence bodies) within the intuitive mind creating more of a balance between our left and right brain functioning. Some of us could even be experiencing a higher understanding of "unity" concepts in our lives from this point forward as we continue to watch the external reality reveal to us deeper insights of what has been going on behind the scene of the "old masculine's curtain".

We are about to step into a new paradigm of which sacred union between the masculine and feminine forces are about to transpire on a "unity code" platform of the creational matrix. This is the architecture of a more spiritually evolved human or ascension timeline of soul mates and so on. However, we are still left to choose and commit to which consciousness force we will serve or engage in - Separation or Unity, Love or Fear. This Unity field is generated alchemically between the "Rod and Staff" principles (i.e. Horizontal meets Vertical, Heaven meets Earth) of the God Creational Forces.

Of course an awakening does not encompass all 7 billion people on the planet at once or in the same manner, but all our consciousnesses impact or create the shared collective reality. Watch TV and we are all focusing or aware of the same things happening around the world. In this way we impact on the whole and our shared reality leaves many different scenarios open – for good or bad - on the world stage through our shared experience.

According to the evolutionary concepts of the Mayan Calendar, this is why it is important that we become more spiritually aware of how our thoughts, words, feelings and actions create reality all around us.

The way we think or use our human consciousness has an impact on the world. Small maybe, per person, but it could happen faster or with greater impact the closer we are to 2012. Although all of this may seem like hard nut concept to grasp for many people, the sooner we come to a realisation of some of it, the sooner we can change our thoughts and actions and the way(s) we create our reality.

Are we prepared for an important choice point or intersection through 2012 - with our future? Are we capable or willing to take this responsibility?

What do you think? It becomes difficult to escape the feeling that the Maya had a mysterious foreknowledge about all this, way before our modern times were born. Why did they suddenly vanish - or leave? Are we currently in the throes of yet another ecological crisis, brought about by human activities, which threatens us with disaster and even survival – like them? Basic resources such as fuel, water, and food are becoming scarce around the world. Many scientists have predicted cataclysmic events due to climate change and pollution over a period of time if we are not careful. On the other hand, we are also experiencing a massive leap in human and environmental technologies. Our world is now meshed together through communications technology and social networks like never before - that act as a global brain where we can easily transmit new ideas or practices instantly across the global information networks. This has happened in less than 20 years – just when our human evolution according to the Mayan Calendar started to accelerate, rapidly.

On reviewing what other authors, researchers and mystics have to say about it, the most interesting part remains the profound implications the Universal Underworld could have on the Earth and the human consciousness. The Maya revered such a peak time of galactic transformation highly and called the time around 2012, The Return of Quetzalcoatl, meaning a return to the light - centre of the universe.

Evolution Is About Consciousness

The Mayan Calendar deals with the relation between time, the universe and the evolution of creation on our planet. The fabric of the universe, the primal Field, was created by energy or consciousness and is held in existence by this consciousness. The entire universe can be understood or seen as the consciousness of the "Creator".

This concept is backed up by so many researchers, scientists, humanitarians and peoples from around the world who have no idea about 2012. Yet they all recognise that people and their consciousness are changing. There is a global, synchronistic and powerful shift sweeping through people as they are opening up to the possibilities of change in new, bigger, more pronounced or faster ways than ever before. The nature of our reality is changing and so are we. So, perhaps we are evolving more than we think?

A Multidimensional Understanding

The shift occurring in consciousness includes a shift to the perspective that life is interrelated. It is about coming to the understanding that we are all a "construct" of Creation. Although the universal "mind" of creation is of course not a consciousness in downright human terms - it relates to the evolution of human consciousness because it is a part of the scheme of creation as well.

The idea that all life carries an aspects or tiny spark of this creation consciousness or "God mind" can explain the accelerated shift we are going through. The Sun has always been worshipped as the giver of life and light on our planet, but as it crosses the galactic equator there is a sacred cross in the sky affecting our human evolution and life.

The Mayans were well aware of this when they constructed their calendars based on cosmic-sun cycle revolutions in relation to life here on Earth. Perhaps we can understand it in a way that we "down here" are inevitably connected or subjected to what is "out there" through universal time.

Even today many Shamans around the world are able to transcend normal consciousness in order to reach higher or expanded states of awareness that stretches beyond their immediate personal sphere or what surrounds them in consciousness through time. The ancient Maya could have received assistance from such ability as well.

If cosmic time seems like "way out there" it may simply be due to our limited perception of time and realities based on a strictly linear or rational mindset. We all know the saying that we can "open our minds" but what does it mean? Are there multidimensional realities?

It is an interesting concept that linear time from a cosmic or universal time perspective is simply a manmade illusion – because as quantum science has been discovering, all of creation is interrelated within a greater universal time scheme. As we pass through 2012 some of us might have a direct experience with this multidimensionality in some way. Eventually, the idea of only living in a 3D dimensional reality or physical body or linear time frame could be seen as a very confined way of perceiving things in life. After all what happens when we die and where do we go?

All we really have is everything in this moment and if anything or a part of us continues after we die, it must be a universal or eternal part of us somewhat similar to an Egyptian afterlife. If we look at life this way perhaps then "Heaven" is not as far away as we think and can incorporate this idea into the "down here", when life sometimes drags along anyway?

CHAPTER 3

ENDING GRAND CYCLES OF TIME

Global Implications of the Shift

If the Shift of the Ages provides us with a link between the Earth and Heaven or the Cosmos, how will this affect us? Will the world see human misery, global famines, war, societal or social breakdowns? Will we realise that we have a saying and decide to institute a new planetary culture or hierarchy based on empathy for humanity, alternative economic systems, sustainable living, ecological design, or an equitable sharing of wealth? If it is up to us to take responsibility for whatever changes that we consider are needed during the Shift – would this move or affect us? If we want to truly understand some of the possibilities in these prophecies, we must go to their source.

Messages from the Mayan Milieu

Carlos Barrios a historian, an anthropologist and investigator who was born in Guatemala has a lot to share about what the legendary 2012 means. In public talks and in private interviews with journalist and healer Stephen McFadden below, Mr. Barrios, has laid out his account of the history and future of the Americas, and the larger world, based on his understanding of the Mayan Calendar and tradition as both an anthropologist and an initiate. He spoke also about the path he sees ahead until the Winter Solstice of 2012.

The range of teachings and insights offered by Mr. Barrios must be considered in the context of his homeland, the Maya of Guatemala, their pyramids, and their calendar which have endured not in a new-age Shambhala of love and light, but in a milieu of treachery, beatings, torture, rape, and murder.

The World Will Not End

Carlos Barrios says he was born into a Spanish family on El Altiplano, the highlands of Guatemala. His home was in Huehuetenango, also the dwelling place of the Maya Mam tribe. With other Maya and indigenous tradition keepers, the Mam carry part of the old ways. They are keepers of time, authorities on remarkable calendars that are ancient, elegant and relevant.

Mr. Barrios studied with traditional elders for 25 years since the age of 19 and says he has also become a Mayan Ajq'ij, a ceremonial priest and spiritual guide of the Eagle Clan.

Years ago, along with his brother, Gerardo, Carlos initiated an investigation into the different Mayan calendars. He studied with many teachers. He says his brother Gerardo interviewed nearly 600 traditional Mayan elders to widen their scope of knowledge.

"Anthropologists visit the temple sites," Mr. Barrios says, "and read the stelae and inscriptions and make up stories about the Maya, but they do not read the signs correctly. It's just their imagination. Other people write about prophecy in the name of the Maya. They say that the world will end in December 2012. The Mayan elders are angry with this. The world will not end. It will be transformed. The indigenous have the calendars, and know how to accurately interpret it, not others."

The Calendars

Mayan comprehension of time, seasons, and cycles has proven itself to be vast and sophisticated. The Maya understand 17 different calendars, some of them charting time accurately over a span of more than ten million years.

The calendar that has steadily drawn global attention since 1987 is called the Tzolk'in or Cholq'ij. Devised ages ago and based on the cycle of the Pleiades, it is still held as sacred.

With the indigenous calendars, native people have kept track of important turning points in history. For example, the day keepers who study the calendars identified an important day in the year One Reed, Ce Acatal as it was called by the Mexica. That was the day when an important ancestor was prophesied to return, "coming like a butterfly."

The One Reed date correlates to the day that Hernando Cortez and his fleet of 11 Spanish galleons arrived from the East at what is today called Vera Cruz, Mexico. When the Spanish ships came toward shore, native people were waiting and watching to see how it would go. The billowing sails of the ships did indeed remind the scouts of butterflies skimming the ocean surface.

In this manner was a new era initiated, an era they had anticipated through their calendars. The Maya termed the new era the Nine Bolomtikus, or Nine Hells of 52 years each. As the nine cycles unfolded, land and freedom were taken from the native people. Disease and disrespect dominated.

What began with the arrival of Cortez, lasted until August 16, 1987 – a date many people recall as Harmonic Convergence. Millions of people took advantage of that date to make ceremony in sacred sites, praying for a smooth transition to a new era: the World of the Fifth Sun.

From that 1987 date until now, Mr. Barrios says, we have been in a time when the right arm of the materialistic world is disappearing, slowly but inexorably. We are at the cusp of the era when peace begins, and people live in harmony with Mother Earth. We are no longer in the World of the Fourth Sun, but we are not yet in the World of the Fifth Sun. This is the time in-between, the time of transition.

As we pass through transition there is a colossal, global convergence of environmental destruction, social chaos, war, and ongoing Earth changes. All this, Mr. Barrios says, was foreseen via the simple, spiral mathematics of the Mayan calendars.

"It will change," Mr. Barrios observes. "Everything will change." He said Mayan Day keepers view the Dec. 21, 2012 date as a rebirth, the start of the World of the Fifth Sun. It will be the start of a new era resulting from — and signified by — the solar meridian crossing the galactic equator, and the earth aligning itself with the centre of the galaxy.

As John Major Jenkins has written in Maya Cosmogenesis 2012; at sunrise on December 21, 2012 - for the first time in 26,000 years - the Sun rises to conjunct the intersection of the Milky Way and the plane of the ecliptic describing in the sky a great cross of stars and planets. This cosmic cross is considered to be an embodiment of the Sacred Tree, The Tree of Life - a tree remembered in all the world's spiritual traditions.

Some observers say this alignment with the heart of the galaxy in 2012 will open a channel for cosmic energy to flow through the earth, cleansing and raising all that dwells upon it to a higher level of vibration. This process has already begun, Mr. Barrios suggested.

"Change is accelerating now, and it will continue to accelerate. If the people of the earth can get to this 2012 date in good shape, without having destroyed too much of the Earth, Mr. Barrios said, we will rise to a new, higher level. But to get there we must transform enormously powerful forces that seek to block the way."

A Picture of the Road Ahead

From his understanding of the Mayan tradition and the calendars, Mr. Barrios has offered a picture of where we are at and what may lie on the road ahead: "Humanity will continue," he contends, "but in a different way. Material structures will change. From this we will have the opportunity to be more human."

- We are living in the most important era of the Mayan calendars and prophecies. All the prophecies of the world, all the traditions, are converging now. There is no time for games. The spiritual ideal of this era is ACTION.

- Things will change, but it is up to the people how difficult or easy it is for the changes to come about.

- The economy now is a fiction. The first five-year stretch of transition — from August 1987 to August 1992 — was the beginning of the destruction of the material world. We have progressed years deeper into the transition phase by now, and many of the so-called sources of financial stability are in fact hollow. The banks are weak. This is a delicate moment for them. They could crash globally if we don't pay attention and we will then be forced to rely on our direct relationship with the Earth for our food and shelter.

- The North and South Poles are both breaking up. The level of the water in the oceans is going to rise. But at the same time, land in the ocean, especially near Cuba, is also going to rise.

A Call for Fusion

As he met with audiences in Santa Fe, Mr. Barrios told a story about the most recent Mayan New Year ceremonies in Guatemala. He said that one respected Mam elder, who lives all year in a solitary mountain cave, journeyed to Chichicastenango to speak with the people at the ceremony.

The elder delivered a simple, direct message. He called for human beings to come together in support of life and light. Right now each person and group is going his or her own way. The elder of the mountains said there is hope if the people of the light can come together and unite in some way.

Reflecting on this, Mr. Barrios explained: "We live in a world of polarity: day and night, man and woman, positive and negative. Light and darkness need each other. They are a balance. Just now the dark side is very strong, and very clear about what they want. They have their vision and their priorities clearly held, and also their hierarchy. They are working in many ways so that we will be unable to connect with the spiral Fifth World in 2012."

"On the light side everyone thinks they are the most important, that their own understandings, or their group's understandings, are the key. There's a diversity of cultures and opinions, so there is competition, diffusion, and no single focus."

As Mr. Barrios sees it, the dark side works to block fusion through denial and materialism. It also works to destroy those who are working with the light to get the Earth to a higher level. They like the energy of the old, declining Fourth World, the materialism. They do not want it to change. They do not want fusion. They want to stay at this level, and are afraid of the next level.

The dark power of the declining Fourth World cannot be destroyed or overpowered. It's too strong and clear for that, and that is the wrong strategy. The dark can only be transformed when confronted with simplicity and open-heartedness. This is what leads to fusion, a key concept for the World of the Fifth Sun.

Mr. Barrios said the emerging era of the Fifth Sun will call attention to a much-overlooked element. Whereas the four traditional elements of earth, air, fire and water have dominated various epochs in the past, there will be a fifth element to reckon with in the time of the Fifth Sun: ether.

The dictionary defines ether as the rarefied element thought to fill the upper regions of space, the Heavens. Ether is a medium that permeates all space and transmits waves of energy in a wide range of frequencies, from cell phones to human auras. What is "ethereal" is related to the regions beyond earth: the heavens.

Ether — the element of the Fifth Sun — is celestial, and lacking in material substance, but is no less real than wood, wind, flame, stone or flesh.

"Within the context of ether there can be a fusion of the polarities," Mr. Barrios said. "No more darkness or light in the people, but an uplifted fusion. But right now the realm of darkness is not interested in this. They are organized to block it. They seek to unbalance the Earth and its environment so we will be unready for the alignment in

2012. We need to work together for peace, and balance with the other side. We need to take care of the Earth that feeds and shelters us. We need to put our entire mind and heart into pursuing unity and fusion now, to confront the other side and preserve life."

To Be Ready for This Moment in History

Mr. Barrios told his audiences in Santa Fe that we are at a critical moment of world history. "We are disturbed," he said. "We can't play anymore. Our planet can be renewed or ravaged. Now is the time to awaken and take action."

"Everyone is needed. You are not here for no reason. Everyone who is here now has an important purpose. This is a hard, but a special time. We have the opportunity for growth, but we must be ready for this moment in history."

Mr. Barrios offered a number of suggestions to help people walk in balance through the years ahead. "The prophesized changes are going to happen," he said, "but our attitude and actions determine how harsh or mild they are."

> *"You must be the change you wish to see in the world."*
>
> *- Mahatma Gandhi*

- We need to act to make changes, and to elect people to represent us who understand and who will take political action to respect the earth. Meditation and spiritual practice are good, but also action.

- It's very important to be clear about who you are, and also about your relation to the Earth.

- Develop yourself according to your own tradition and the call of your heart. But remember to respect differences and strive for unity.

- Eat wisely. A lot of food is corrupt in either subtle or gross ways. Pay attention to what you are taking into your body.

- Learn to preserve food, and to conserve energy.

- Learn some good breathing techniques, so you have mastery of your breath..

- We live in a world of energy. An important task at this time is to learn to sense or see the energy of everyone and everything: people, plants, animals. This becomes increasingly important as we draw close to the World of the Fifth Sun, for it is associated with the element ether – the realm where energy lives and weaves.

- Go to the sacred places of the earth to pray for peace, and respect for the Earth which gives us our food, clothing, and shelter. We need to reactivate the energy of these sacred places. That is our work.

- One simple but effective prayer technique is to light a white or baby-blue coloured candle. Think a moment in peace. Speak your intention to the flame and send the light of it on to the leaders who have the power to make war or peace.

We Have Work to Do

According to Mr. Barrios this is a crucially important moment for humanity, and for earth. Each person is important. If you have incarnated into this era, you have spiritual work to do balancing the planet.

He said the elders have opened the doors so that other races can come to the Mayan world to receive the tradition. The Maya have long appreciated and respected that there are other colours, other races, and other spiritual systems. "They know," he said, "that the destiny of the Mayan world is related to the destiny of the whole world."

"The greatest wisdom is in simplicity," Mr. Barrios advised before leaving Santa Fe. "Love, respect, tolerance, sharing, gratitude, forgiveness. It's not complex or elaborate. The real knowledge is free. It's encoded in your DNA. All you need is within you. Great teachers have said that from the beginning. Find your heart, and you will find your way."

Universal Prophecy Themes

If we find ourselves in a time of trials, it is a sacred testing, or a great crossing, maybe. We must find the ways to live in harmony and peacefully coexist within our biosphere; our survival depends upon our awakening to the interdependence of life. By the day, it is becoming more evident that business as usual is the blind path, so we must learn to navigate with greater insight and ability to work together and cooperate in new ways.

Regardless of any specific calendar end date, as the indigenous elders confirm, the signs of the times are manifesting everywhere - both inside of us and in our world. We will all feel the intensity in our lives as we progress through these changing times, somehow. Whether we look at the global economic crisis, the environmental and social problems, human overpopulation, poverty, homelessness, disease, starvation, wars etc. it is clear we are living in significant times. This moment in our evolution as a species has never existed before and it is clear in the way, it is right now.

This living prophecy is not about one day where a light is suddenly being turned on in the world rather this is about a process we are in NOW, that will keep steadily unfolding. The essence of this time of prophecy is a time of crossroads; a time of shifting cycles where the old world age and its separation-based mentality is in the process of dying and transforming into a new world that will live and create through our hearts.

We are in a great in between time, experiencing a tension of the opposites playing out between contracted forces of fear and darkness that seek to have false power by manipulating life with a mind founded in duality, and the expansive forces of possibility and light that seek to be empowered by living in harmony with life, and surrendering our hearts to the mystery school of life that unites us all.

An important theme of this living prophecy reminds us that "it is always darkest before the dawn." In other words, it is said that if we can navigate through these times of immense challenge and crisis, the eventual renewal of the world age or cycles will bring a gradual return to wisdom and balance.

The theme of this return appears in many different prophecies: Return to Balance; by consciously unifying matter and spirit; Return of the Divine Feminine. Return of the Christ Consciousness; Return of the feathered serpent, the snake and the bird unifying the lower, primal and higher, celestial nature within us. The Return of the Wisdom of the Ancestors or Star Beings; The Return to Living In Harmony with Nature; The Sacred Hoop of All Nations can be Whole again; the New Jerusalem emerging, the Rainbow Warriors arising; Returning to Awareness of our True Nature etc. Could it really be?

In simple terms, yes, it seems there is no other quest but to awaken to our human potential. The focus of return or awakening can bring a potential rebirth into a new sense of wholeness in consciousness. Essentially, the idea is that we have been in a long era dominated by the linear, rational mind split off from the body and the heart, entranced by linear time, old scientific paradigms, blind materialism, egotistical competition, dualistic mentality of right and wrong, and a fear-based sense of separation and greed resulting from a deep, societal, entrenched disconnection from the sacred oneness of life.

As we gradually shift gears out of the old paradigm described above, we can eventually move into an era that will be increasingly guided by the heart and intuition integrated with the mind and body, attuned to the eternal dimension of being, supportive of new quantum sciences, living in harmony with nature and in touch with the spiritual, invisible dimensions and forces, dedicated to authentic cooperation, tolerant of diversity and holding space for multiple viewpoints, guided by conscious awareness of our precious interdependence, desiring to serve the whole.

Although of course, we are not all prepared to do so, do all of that, or do it at the same time, each of us can start somehow, somewhere and according to Mr. Barrios has a role to play. Carlos Barrios, therefore, offers a number of suggestions to help people walk in balance through the years ahead:

"We have the opportunity for growth, but we must be ready for this moment to act, make the necessary changes, and elect people to represent us who understand and who will take political action to respect the Earth. Meditation and spiritual practice are also good. It is very important to be clear about who you are, and also about your relation to the Earth. Develop yourself according to your own tradition and the call of your heart. Remember to respect differences, and strive for unity. Eat wisely. Pay attention to what you are taking into your body. Learn to preserve food, and to conserve energy. Learn some good breathing techniques, so you have mastery of your breath. Be clear. Follow a tradition with great roots. It is not important what tradition, your heart will tell you, but it must have great roots. We live in a world of energy. An important task at this time is to learn to sense or see the energy of everyone and everything; people, plants, animals. This becomes increasingly important as we draw close to the next Sun cycle, for it is associated with the fifth element, ether - the realm where energy lives and weaves. Go to the sacred places of the Earth, pray for peace, and have respect for the Earth which gives us our food, clothing, and shelter."

"The world will not end. It will be transformed.....Change is accelerating now, and it will continue to accelerate...

...If the people of the earth can get trough this 2012 date in good shape, without having destroyed too much of the Earth, we will rise to a new, higher level...

...Humanity will continue, but in a different way. Material structures will change. From this we will have the opportunity to be more human..."

- Carlos Barrios, from the Eagle Clan of the Maya of Guatemala

The Shift of the Ages Is All Around Us

As we are preparing for and going through some of the greatest accelerated changes our world or civilization has known through and beyond 2012, the signs that old cycles are ending and a Great Shift is upon us are all around us.

Although the cause of these endings or changes for the most part might be associated with other things to many people, the implications of an ongoing Shift are nevertheless here, perhaps vaster than we can imagine or dare to think with regard to our lives, societies and cultures in future. In the years to come the changes could become unprecedented and the ways they affect the way we have seen ourselves and our ways of living in the world.

Although the Mayan Calendar form the basis of this understanding of the biggest change challenge of our time, we do not need to explore ancient prophecies or predictions to see that profound evolutionary challenges or changes are brewing underneath the surface or could reach a boiling point.

As we take a look at the news about what goes on in our world, from earth changes and quakes, to upheavals in our political, religious, economical and societal systems, we can see that world events are escalating or happening faster as if time or evolution was indeed speeding up almost by itself while we try to keep up with it all.

Individual Implications of the Shift

If we look deeper, we can see that it might reflect something that also goes on within us. It is not just our world, but our individual life experiences that are subject to these changes as well. Many people are losing jobs, change residence or careers, relationships, life paths etc, in increasing numbers or at an increasing pace. It is through whatever we are going through – that we are becoming more aware of the unusualness of our times.

As the Shift of the Ages continues to unfold, it is important that we understand what it is trying to teach us through the situations and things that we need to adapt or adjust to. In order to do so, it is essential that we keep our eyes open and listen to the hunches and feelings we have inside of us. The effects that we see in the world are all, but clues to what are changing within and without. Although it is clearly a challenging time for many people around the world, it can also be one of the most exciting times to be alive...

If we pay attention, according to the prophecies held by the Maya and other indigenous cultures, we have the potential now to integrate spirituality, wisdom, indigenous shamanism and eastern mysticism, with the physicality of our lives. We may also see new scientific breakthroughs. All of it is supposed to happen in order to lead us to a new point in our understanding of the link between the physical reality and our spiritual potential.

By seizing the moment, we can realise more of our potential or power. Therefore, it is important that we take our attention off of the 2012 date as being some linear point in time triggering some event that is going to determine our fate. The reality is, we are on a journey, right now, through the shifting world (age cycles). Where-ever we find ourselves, we have to understand this is a process, not a product that will arrive on a particular day and time. It is obvious, that focusing on future projections, imaginations or problems is a distraction to this moment and its opportunities of empowerment. While maintaining awareness of the larger context of this time of prophecy, we must understand, here and now is where our life is playing out.

Also, in a broader sense, if we contemplate the theme that is being associated with the 2012 prophecy, we can learn a lot by looking into the original (biblical) meaning of the word Apocalypse: Revealing of that which has been hidden; lifting of the veil; uncovering. We can see this in our world as covers are being lifted, veils are thinning, and we are being shown many new things that have previously been hidden. The corruption or insanity of those in positions of power in the world, or the chaos within our selves is being rapidly exposed, just as is the profound and untapped depths of beauty and wisdom which stem from living within our hearts could be revealing itself.

The years surrounding 2012 could therefore bring a time of positive transformation and the opening to whole new ways of life - a kind of bridge perhaps between what we think we will experience and what we could experience if we care to see beyond the physical by bridging it with the spiritual.

Are We Going with or Against the Waves of the Great Shift?

As we are leaving an old cycle behind and reaching for a new one, the spiritual changes could become predominant, particularly around the galactic alignment. Will we be forced to let go of past assumptions, religious beliefs or a faith in a strict rational-materialism through sciences, endless economic growth etc.? We are certainly entering a (spiritual) cycle of growth that will be very different from the (material) cycle we have become used to. Some say a spiritual revolution is upon us although many of us may not be cognizant of it.

Ignoring or actively resisting the impetus for change (arriving in consciousness) might not result in immediate problems, but If we do not make necessary adjustments in our individual lives and through our collective body it might turn out more difficult than it needs to be when waves of change, be it spiritual, personal, economical, environmental, societal or... arrive.... for a reason... and affect us in some way.

The question remains how we can use our present understanding to deal with the order of the day and learn to adapt. Our human mind, according to the Mayan Calendar, may be challenged based on our past belief systems, coping mechanisms and conditionings based on our consciousness' development during the prior Mayan underworlds. The way we learn to handle the new waves that are coming will impact whether or how, we will be able to move forward into a more positive future of increased potential and growth.

When the Mayan Calendar reaches its end and time flips over into a new cycle, our old human consciousness might start to flip over too. Questions then arise about our past, present and future; will we start to work with each other for a better, more peaceful, prosperous or sustainable world or will lack of temperance, our human ego or spiritual progress cause further destruction, devastation or chaos within human life or the planet's eco-systems? Either way, we can threaten or increase the survival abilities of the future lives of our children and grand children. What we do now will become decisive for future generations on our planet, because we are the ones currently inhabiting it. In other words the planet or the future does not belong to us, but depends on our conscious ability to work with it instead of going against it. Will we realize this – or will it be too late?

Perhaps the real question we need to ask is how can we develop our human consciousness or expand our awareness in order to grow as human beings during our presence here. Without daily awareness of what goes on not just in our lives, but in other people's lives as well, change could mean serious trouble, when we as humans have a tendency to fight for our limited opinions or personal agendas or avoid change all together by hanging onto what we already know out of security or fear of what lies "on the other side".

Swimming upstream against an increased flow of changing events, could turn the tide against us – and flow over into unnecessary rivulets of hardship, where distorted water ripples or waves come splashing or crashing on our shores, none of which we have chosen, but would need to swim against.

If we chose to go with the flow, rather than against the sea of change, we have an unprecedented opportunity to become more aware, evolve and progress through life in a spiritual way. Although this does not mean we will all of a sudden "go spiritual" – we nevertheless have more of an opportunity to awaken right now.

It is really a matter of our level of openness and willingness how much we are able to explore or integrate more of the spiritual with the physical or worldly side of life. If or when this happens, however, it is not just because it is handed down to us from "up there", but because we chose to embark on a personally transformative journey that affects the planet as well.

Can We Surrender to the Shift? When? How?

What will happen as cosmic time accelerates? Will it trigger a quickening of the evolution of our human consciousness? Will we sense it? If so, how will we know or start working with it?

Of course, to some, the ideas of a 2012 Shift may seem farfetched, as an otherworldly spiritual event in a context that is off the wall. Others may welcome it as a new idea or exciting piece of information that could trigger further "aha-experiences" and for yet others it could cause quite a stir they never thought or dreamed would happen to them or their lives. We all see the world from different personal perspectives so going through something as immense as this could trigger all kinds of reactions from people around the world. What we need to do perhaps is take out some time to study or think about it ourselves and how it might relate to our lives. Eventually, we are all here to learn from this life whether we agree to it or not and the universe probably did not put us here by total accident – meaning life or God has a tendency to throw us things that we need to learn on our pathway through life's unpredictability. Human life is more often than not a mix of sunshine and clouds – and very few of us go through only fair weather or experience "clear blue skies" forever. Whatever we do or do not do about our life experience and what we go through here, there is always a strong element of free will and choice as to how we let life colour our human vision. Often we need to think again, look further or deeper to figure out what is really happening with us in life. To figure it out we may have to retreat or spur into action contrary to what we might initially want. Most of us do not like the idea of change because it threatens our ego's tendency to protect itself and our personality's reluctance to stretch itself and avoid pain. Even when or if we need to make a change the most – resistance has a tendency to seriously get in the way - leaving us stuck in the same ways. Perhaps this is why so much of human life often stays the same. If we are fragile human beings, we would rather stay where we are or with what we have, because how do we know what we will get or where we will be if we risk something under our noses?

How often do we seek or realise the meaningfulness in life experience that are like an uphill struggle, even when this is where or when we often learn the most because it teaches us that we are stronger than we thought or that we can move forward against all odds? When life is like an easy ride down the hill again – there is not that same incentive to strengthen the psychic muscles or figure out how to do something or go somewhere else. Moving beyond ourselves depends on our visionary willingness and experiential focus. If we are to change it is often because we are pushed to grow or evolve where it matters not if we hold onto our past frame of reference, but if we allow ourselves to grow a little each day perhaps it becomes easier to make a big change tomorrow?

Life is not "black or white" there are many colours or nuances in between. If we can find a sacred middle way on our spiritual path of non-resistance it often becomes easier than going to some kind of extremity. Perhaps this is what these "in between times" are really about if we are to use and go with rather than against them. When or where old ways are giving way to new ones we need to allow them to be born inside of us, first.

The spectrum of life is as varied and diverse as there are people on the planet and more, because we often experience or integrate similar life experiences very differently. The way we get on with life of course depends on who we are, where we have been, where we want to go and our ability to cope with what lies in front of us. Similarly, there is not one path, but many options as we progress on life's journey.

It all depends on our frame of reference and openness about the future, how far we are willing to go – with a change.

Often times we need to see other people with traits or performances that we can then use as role models, before we go ahead with something we have not tried before ourselves. New ideas about this or that in life are often first or initially discarded or ridiculed – before they eventually become accepted or applied by the masses after a number of us reach a critical mass. From then on it is every man's everyday knowledge. We have seen this with anything from personal trends to social changes and worldly realisations. Once the Earth on which we live was considered flat until explorers who travelled the seven seas managed to dismiss that and what was once a fact became disapproved. Now, obviously we are standing at a whole new threshold.

Enigmas or new spiritual realisations coming from out of nowhere do not necessarily mean we are going crazy. How can we be willing to recognize, understand or know anything in our human lives related or connected to what goes on in the universe or cosmos – if we cannot see it because the vision is covered by the Earth's moving clouds?

How many of us look at the stars in the night sky and wonder? Where do I/we come from? Is life on Earth really all there is?

Although these are timeless existential questions, they have a lot to do with change processes during the Shift because if we do not question our beliefs about what we consider the truth in relation to who we think we are – how can we ever turn in another direction or "convert" to the idea of a "Big Spirit" as a kind of divine intelligence that penetrates the universe? Even if it does not matter much to us personally how the Earth or we were created, would it not be nice to know whether we as Earthlings arose out of more than stardust that settled by coincidence here on Earth?

And what about aliens – do they exist? If you keep reading, you probably already ponder or know the answer to some of these questions. No matter where you are on a spiritual belief spectrum the questions you ask matter to the answers you get. Whether we are consciously aware of it or not our spiritual understanding or awareness is important because it allows us to explore life and with it the Shift rather than ignore, defy or go against ideas associated with it. The most important perhaps is how open, willing or committed we are to "turn another page".

Those who have studied something "spiritual" or been involved with "New Age", whatever that means to you, may find spiritual ideas or material subjects more digestive or familiar, but it does not exclude people who find this less accessible. Oftentimes "non-believers" who totally disregard these things, practice spirituality daily by simply following their heart knowing what is true. It is all a matter of putting fixed ideas or beliefs about spirituality to rest and rely on sources that resonate with our inner guidance in spite of whatever reference.

When we are ready for something new or more "spiritual" it often appears as if by some kind of magic by revealing what it has to reveal to us at exactly the right time at right place through acts of synchronicity. This is because we are supposed to get to know something to evolve or move from whomever or where ever we were to someone, somewhere next. Essentially this is what we all "came here for"; to learn from life as human beings in order to evolve spiritually. For some it takes longer than it does for others to finally get it or share it. In a way or in a kind of nutshell you could say that this is what the Great Shift is about. Through personal first hand experience, comes knowledge and insight that can be transformed to spiritual wisdom when we are ready to receive it through the heart.

The Shift has arrived for a reason – and if not from Heaven – it will most certainly kick up a lot of "old dust". To some this looks like we are going to Hell, but we have been on this invisible, accelerated course of awakening for quite sometime from the late 1980s and will continue to do so through 2012 and what lies beyond. What we have seen as a result rather; is a growing collective interest in "spiritual stuff" and how it relates to daily life through holistic health, self growth, self help books, new religious waves and social movements as we go green or "outside the box". Anything that diverts slightly from old ways of thinking, being or doing things in the past means we are in some way evaluating our personal stand points about certain things. In a way this trend is already reshaping our world, although perhaps initially with small baby steps, but they are creating a new alternative, collective, global atmosphere that allows new ideas and ways of life to be able to walk and run their own course with time.

These shifts happen both visibly and invisibly when there is a sudden change of mind or understanding, including changes in ideas about how the human mind-body and spirit relate to what is outside of our present or immediate sphere of influence.

How we think or what we think affect not just our personal well-being, but our human relationships and the Earth's environment etc. We are finally beginning to understand we are not separate from the Earth's biosphere. We are it and what we put into the environment has a direct impact on us through "climate" change.

Many questions are now being asked about the sustainability of our old lifestyles. We are questioning our reliance on dirty fossil fuels when renewable, free or sustainable, green energy and other alternatives energy source already exist on the planet. Why are we then burning it all off? Can we become (more) independent of these fuels by taking some flatfoot ideas one step further before the gasoline runs out or reaches a sky high cost? It is all part of where we are going in consciousness while the planet is also "shifting".

With or without many people necessarily understanding how this is associated with a Shift in our conscious human evolution, it nevertheless shows us a growing urge, interest or need to move in the direction of a more sustainable planetary future.

The more alternative or "spiritual" we become, the greater perhaps the desire to embrace such new concepts or ideas. Who is right and who is wrong about "going in a new direction" is pointless, really. Our survival and progression as a species depend on our ability to manage our environment and earthly resources better when billions of people struggle to survive. We cannot leave these problems till a day after tomorrow, far from it.

We need to exercise our human cupid and psyche and to find new solutions before the planet reaches a boiling point. No matter our understanding of interrelationship with the planetary "sphere" it is important that we start doing something by working together to secure our future across the globe in spite of our religious, racial, cultural, societal or other differences. They have endured long enough and if we are to change we need to turn away from "I" and "we" to "you" and "us". We can no longer afford to blame our neighbours or another (party) for what is going on or going wrong. Global solutions are greatly needed to face the challenging issues we are facing during these times. They concern us all. If we do not we risk splitting our selves or the world further apart, where we are running a serious risk of compromising our future without a safety net. The Shift, more than anything, will perhaps teach us that......

How we learn or whether we start learning it, depends on our ability to go for and with it rather than contrary to or against it. If we do not realise the importance of working together – a "sea of global change" could throw us off the ship or cast us ashore. No matter what, we need to consider there is a cause and effect and our past affects the present which in turn shapes the future in retrospect. If the universe is currently giving us a big hint or wake up call it is by showing us that there is no economic growth model that can free us from addressing our planetary problems.

It derives from the industrialisation ideas that arose with scientific exploration through the prior Mayan Underworlds. We have already seen one Roman Empire collapse once. Is the Western hemisphere with its rational-material predominance going next with the East hanging in its threads?

Planetary consciousness-raising is just one way we can work with the Shift. We all need to take further responsible actions.

What about the big car? Can it be replaced with a bicycle perhaps? Can we grow fruit and veggies locally instead of buying the imported and transported ones? What about the cocoa beans – are they harvested by children who work as slaves? Is the tuna – dolphin free? Yeah, a new "wave" has arrived alright, whether we care to work with it or not. It is changing the; I don't care, or what does it matter to me replaced, with an immediate or growing concern for the local or global environment because the world is obviously "shrinking".

Anything from news coverage of crazy weather disruptions to volcanic eruptions can make us think twice or become more aware of the seemingly mindless way we have been thinking or acting before. We can no longer afford to think only about today, because today becomes tomorrow and by the time it does, it is too late if no one knows or cares about that. We are all breathing the consciousness or unconsciousness of life and the way we do alters "the air".

Those with fixed or mixed feelings about such a Shift may soon need to think again or revamp themselves because those who are most reluctant to change could suffer the most if they are not prepared.

We can reject, embrace, like, love or fear the ideas about the Shift, but going against what will happen is only bound to make matters worse. It already hovers in the ethers and depends of course on how we chose to work with the time changing tides. In order to avoid a point break perhaps we need to asses the waves that are crashing on our personal shores?

The world is round...and the place

... which may seem like the end...

... may also be the beginning.

~ Ivy Baker Pries

The Shift - A Potential Leap in Human Consciousness

Whether we are filled with a sort of expectancy, excitement or anxiety about what the threshold around 2012 evokes, any of us who are looking to the Gregorian calendar for guidance to align with a new time, will be greatly disappointed. Very little is provided there, other than a linear segmentation of time into years, months and days. We have already discussed that. In contrast, the workings of a sacred calendar, such as the one created by the ancient Maya, provide a wealth of symbolism able to empower its adherents with greater insight, energy and clarity...for changing times.

The Mayan Calendar outlines the evolutionary pathway that is laid out for us. The question is how we work with and learn from it. What have we gained so far from the past's experiences, and how will we take ourselves to the next step?

From a birds' eye, it may seem like we still have one foot in the old world or reality of the previous time cycle, but when we look closer our feet are already stepping onto the beach that meets a new sea. If it does not feel this way, it could be because we are still standing on an old beach waiting for the old waves to wash away the sand in our footsteps. Most of us have gotten so used to the Gregorian Calendar's linear measurement of time, that we pay no attention to what really goes on now and simply keep running down the same track day in and day out, as if time and life was progressing of itself in an unconscious fashion within our minds without any conscious efforts or decisions of ours. In this way, the linear time perspective has a tendency to control our lives, rather than open our minds to the many spiritual opportunities or pathways that are made or becoming available to us. We have a choice now to stop in our tracks if we are willing to take a pause in the middle of the grander scheme of things...

How can you be better able to feel and pay attention to what is really going on in your life? Do you know or recognise what goes on inside? Would that make you turn your life in a different direction? Or change your course of action? If you have been directed by linear time and suddenly turned that around – it changes your perspective because the focus would shift from time chased to the time chaser. You could be stepping into cyclical time where life starts to make more sense and a deeper part of you attaches a different meaning to life's events. Do you see?

Our life was never meant to be ruled by time alone in the way we have let time control us, everyday. Worldly life is meant to also be an inner reflection where we are consuming time instead of being consumed by time. Time spent on something is perhaps not as important as how we experience what we spent it on while we were doing it - do you see? This is what cyclical time is about....

If you are living life outside-in, instead of inside-out or your head rules your feelings how can time become other than a disallowing experience where you are not being in the right place at the right time. Linear time runs down our veins and when they become devoid of bloody life force it is time to change perception and find out how the content fits the context. There is nothing wrong with time passing. The problem is when we let it pass or slip away through our fingers. This is why many of us may have been feeling a lack of spiritual interest or significance – because the spiritual connection that keeps us truly animated and alive no longer runs through our life blood.

If we were more in "real time" – our spiritual selves would become better integrated with the worldly life revealing more of life's precious gifts at the right place at the right time. Somewhere along the line we officially lost it with the Gregorian calendar's utilitarian approach to time, because time spent there restricts our ability to manoeuvre in "real time" in relation to a grander scheme or plan. The Maya knew all about this. They were obsessed with time because they knew how it affected human life here on Earth.

Where do these times lead us? Does time take us to where we need or want to go? As The Shift goes on it allows us to reassess or adjust our relationship with time. Our present life path is where old and new intersects. It has particular importance or interest during this time which contains many new signals. What goes on in the "mini-universe" relates to the "omniverse"- as in micro-macro - and allow us to merge our "selves" or "spaces" with the right time - right place during the Shift that is arriving in incremental, yet steady pieces.

We can either choose to grow and evolve during or with this time or struggle to adjust through changing times. Although this does not happen overnight, as the last grains of the old cycle spiral through "the hourglass of 2012" and beyond, it may indeed start to feel more like we are actually "doing or making it" over the coming years.

If we were able to look back upon the time of the Shift from a point in the future through the eyes of our children or grandchildren - we could possibly see how far then we had travelled in our human consciousness and evolution over a fairly short time span – surrounding 2012.

The question remains; what does it mean and what will we do about the quickening taking place in relation to our future? If it cannot be stopped, can we choose the access point and how to enter it? When? Where? How?

> *"To transform itself in us, the future enters into us long before it happens."*
> *- Rilke*

PART 2

BRIDGING THE OLD AND NEW CYCLE

Photo (previous page)

Crop circle formation, Stonehenge, England, 2010

Image Credit: Lucy Pringle

CHAPTER 4

THE TRANSFORMATION OF OUR WORLD

Mystical Interpretations of the Shift

As the Sun sets on the old cycle and we move through 2012 and a new sun cycle begins, it is interesting to share with you what some wisdom teachings have to say on this topic.

As with all cycles, transformation does not begin or end on one day; instead it develops or unfolds over a period of time. Even if we wish for a magical moment of time, where the world is instantly transformed for the better perhaps – there is not one act or one secret that changes everything. Even in fairy tales or myths we often see that when heroes find the magical talisman, their journey does not end there, rather it has just begun!

The unit of measure that affects everything on Earth is time. Mayan wisdom teachings state that time is the keeper and ruler of our universe; it is cyclical and it moves in (expanding - constricting) circles. We are in a constricting circle now before time starts to expand again on the other side of 2012.

When we look at the beginning of the previous Sun cycle during the period of 3000 – 3115 B.C, remarkable occurrences were happening around our planet. The first dynasty of Egypt was born, the first records of written language were recorded in Mesopotamia, the Hindu cycle of Kali Yuga began and the Sumerian civilization developed the first monarchy. These civilizations mark a majority of the history we use as reference points even to this day.

The Mayan prophecies connected this 5,000-year cycle, which began around 3100 B.C. and culminates in 2012 with the 10 troubled times that they referred to as the 9 Hell Cycles. This cycle indicated that humanity would undergo a long period of fighting and struggling with what we can call our shadow self (the Maya described it as the dark side of the ancestral underworld).

As we near the end of this 5,000-year period, we are preparing for a shift into a whole new cycle where human life and evolution will be experienced differently. The Mayan prophecies highlight a potential 26-year cycle on either side of 2012 as the most important period in our human evolution which will define the next human evolution and civilization cycle. As we are near the end of the old period, Dr. Jose Arguellas, who has studied the Mayan/Toltec calendars in depth, lists three important dates: 1987, 1999 with 2012, the culmination. The first date is known as the Harmonic Convergence, which began August 16, 1987 and would be felt on the Earth in conjunction with our sun's 26,000-year precession of the Equinoxes cycle.

The significance of the Harmonic Convergence was reputed to stir an energetic quickening that would stimulate movement back to a more balanced spiritual centre, assisting in releasing the shadows and delusions created during the 9 Hell Cycles described by the Maya. Their wisdom teachings describe the specific point in time around 2012 as a preparation period to learn to open the human heart and mind to a new level of consciousness in order to release the delusions or mindsets held in place by the human "shadow" or "lower self" of the ego-mind.

In 1999 we reached the midpoint of this cycle and began a 13-year period preparing us for the transformational time highlighted on December 21, 2012. Many say that the most important work to be done during this period is yet to come with self-introspection coupled with the release of old ways of being that might not be working after the Shift because the new time cycle is related to a "new consciousness field".

One of the most challenging psychological issues we are struggling with in the world, today, is the human ego's need for control. Our biggest problem with it, is that we are still trying to "logic this life out", thinking in ways of how we can control or manipulate even the Earth and its environment with lower thoughts. If we live life in that frame of mind, we may experience fear of the unknown in relation to our personal, human survival needs of control.

As we move through the 2012 transformation of cycles, we are transcending into a new era with a new kind of consciousness. This, in turn, will alter our perceptions of how we view ourselves and the world we live in. As with any period of change, it begins with conflict and chaos, within and/or without as we are seeing in our world.

The time in between two cycles is giving us the opportunity to deal with our shadow self or transcend the past in order move into a higher level or new frequency of life. A spiritual change or shift in consciousness often begins, with the illusions of the ego being exposed, so that we can confront life without the limitations of our ego identity masks we have been dragging along with us for a long time.

As this is revealed, issues surface in people and parts of humanity that have a great need for control when old or outdated ways of thinking or living no longer work the same way we have been used to in the past.

Ultimately, this turning point in time is calling out to each of us to grow (beyond our fears) with courage, compassion and creative determination - all of which can emerge from connecting to our hearts. When we look at the state of the world affairs from a rational, linear mindset it is easy to see terrifying scenarios of doom or endless cause for alarm. But the energies of fear and panic will get us nowhere. Instead they can lead to stress, sickness, overwhelm, negativity, depression and escapism. Whereas the old paradigm model used fear to motivate us, the new paradigm is based on allowing our love for life to excite and inspire us.

Is it not better to change willingly now while we still have the opportunity to level this process out before it might go out of hand? Understanding the Mayan Calendar and the 5,000-year cycle, we are leaving behind, means surrendering the shadows of our fear based ego mind. If we can convince ourselves that we are not separate from God and allow universal wisdom to enter our minds, the process is bound to become and easier one. At least ancient, esoteric teachings tell us, that the heart is the centre of true wisdom, and the key to unlocking the divine or spiritual mystery that is ours by birth right.

As we also move closer to the cosmic heart of the galaxy, through the galactic alignment of 2012, the linear logic of the mind will not assist in this shifting time process. Only if we open our hearts can we prepare for a conscious connection or reconnection with God. This is a hermetic axiom referred to as "as above, so below, as within, so without". In other words; what exists in the universe above is reflected within us below. The heart of course is the centre of our bodily universe – just as the Sun is the heart of the solar system and the Galactic Centre the cosmic heart of the galaxy.

We have (chosen) this moment to be here in time. We did not come here all by accident, but got a front-row with a VIP pass to participate in the Shift of the Ages, felt by both Heaven and Earth. Do you sometimes wonder why you are here or what you came here to do? If so - this might be part of the reason. While you think about preparing for such an event - even if you have no clear idea of what that means to you or how it will affect your life - it is already unfolding in and around you.

What you do each day, willingly or unwillingly, brings energy, action or momentum to the Shift. How often do we not see ourselves as separate from the whole or long for one breath-taking moment instead of realizing that this life is all we have now, and it is precious. Each day through 2012 and beyond is a sacred opportunity to open our hearts and shift into the spiritual beings that we are at a core level. It is a changing moment in time that could take our breaths away – if we allow it.

Of course nobody can promise it is going to be an easy ride or a walk in the park. It could become a time where we have to face uncertainty and self-doubt. After all we are in the culmination of a 26,000-year cycle of change and so we are forced to change more quickly, so many things could feel uneasy. The question of course is when or how much we will be able to stand or allow - before things get messy? Human nature has its obvious weaknesses and fragilities, particularly when we are under some form of "pressure".

Some of us have families that would have a complete meltdown in the morning if we are out of a certain breakfast cereal, or change the coffee brand and here we are, talking about an energetic Shift or change that could affect over 6 billion people on the entire planet along with its animals, flora and fauna. In a way it is both a personal family and global affair – whether we believe in 2012 - end date or not.

So, how can we take what we have learned through the past cycle with all its destructive, war-like behaviours and apply it in a different way - to a new level within our evolving human civilisation? Is it possible? Well, the graduation date is upon us, and it seems our senior years are all revolving around 2012.

Even if it seems overwhelming, this is our time to look beyond what appears to be of the dark and see the light that is always present in all of us. If we find ourselves in a shadow time, or a time of releasing, which has been created by our doubts and the collective human fears, we need to see through the tunnel. If the Earth is transcending into a new age, so are we, bit by bit, soul by soul. This time period will never come again in this way and somehow, whether we are aware of it or not, it might affect us all. In fact it is already happening.

> *"I am only one, but I AM ONE. I cannot do it all, but I will not refuse to do what I can and what I can do, by the grace of the Creator, I will do."*
>
> *- Edward Everett Hale*

Preparing for the Great Shift

The Shift as a rare, cosmic- galactic alignment means an increased cosmic light pulse is emanating straight towards us from the galactic centre through our Sun to the Earth. Like an invisible force field - the spiritual energy associated with it - will permeate the Earth's quantum field during the years surrounding 2012 and beyond. As the energy arrives in waves or unprecedented amounts and we spiral through the end of the old cosmic sun cycle - it could mean changes not just for our planet or world, but also to the human mind and body.

According to some Hindu spiritual traditions this solar power is associated with the Kundalini – which is defined as a vital energy force running through our bodies. Although it is not completely understood – at least scientifically - how this energy flows through the human mind- body system - it is a universal truth that our body runs on a vital energy which keeps us alive. The Chinese call this Qi or Chi for aeons. The Indians call it Prana. The ancient Polynesians or Hawaiians called it Ki.

The Shift is bound to change or affect this Ki or life force as universal or cosmic energy potentially works its way "outside-in" and our bodies interact with it "inside-out" when more invisible, universal or photon energy suddenly drops down from the cosmos. Some say that cosmic energy changes the yin-yang polarity of the mind and body and could cause sudden or gradual, spiritual transformation awakening that will affect our human evolution in consciousness. Of course no one knows exactly how, but as we are going through the Shift over years in linear time, it could trigger many different things within us in ways that are hard to predict. Much will depend on our ability to adjust to it and work with what happens in our lives, accordingly.

Will it be easy or difficult, sudden or gradual? Because we find ourselves at very different ends of life's spectrum, there is no easy answer to that question. The ways we are going to respond will vary greatly. Perhaps what is most likely to occur, is that it will trigger a gradual rediscovery of our true self and interconnectedness with the rest of the cosmos. After all this understanding is the foundation of the Mayan Calendar's Universal Underworld in the first place.

Since August 1999 we have been given an opportunity to look at our inner self as it is reflected in our outer world, allowing us to make the necessary choices or changes that are needed to transform our world or heal our Earth in relation to the cosmos.

Although many characteristics of the New Human or New Earth may look very similar to the old one on a surface level, it might nevertheless "feel" or "seem" very different, the further we move along the higher frequency phases of the Shift and the ways we globally manage to interact with it as we are being exposed to the cosmic pulse or waves emanating from the Galactic Centre.

The effect of such a rare galactic alignment possibly marks the birth of a new civilization, based on a more enlightened humanitarianism, new technologies or forms of energy. Somehow, possibly along the way, as the intuitive faculties of humanity awaken, we will explore or embrace a more spiritual way of life with a deep regeneration or renewal for humanity together with an individual search for deeper spiritual meaning and truths of life.

Can All This Really Be Happening?

If you are wondering whether or how all of this can really be true or happening, it is only natural. When we look at things at first glance it is always a good idea to contemplate them further, before we agree or discard whatever is being presented. When we look at the face of the world it may at first glance seem as if nothing is really changing at all, but what if we look more closely? The best way to detect the Shift is perhaps by using our "antennas" as we pay attention to what goes on with the world around us - using our sensitivity, intuition and common sense.

If we look beyond surface appearances, it is not difficult to see that something unusual is occurring in our world, where things are not right the way they used to be......

What about the sudden climate changes we are seeing? Why we are seeing and experiencing so many Earth quakes, volcanic eruptions, tsunamis and the like? Why are hidden issues surfacing or being exposed worldwide? Why do governments, established institutions and leaders try to tighten a grip hold? Are they afraid of losing power or control? Why do so many people seem to struggle? Why is there so much chaos or confusion in the world?

Whether we can understand what goes on with the world or not, find ourselves in apathy or able to cope - asking deeper questions gives us an incentive to find some real answers. Rarely do we get real clarification by reading or watching the global news or listening to our leaders when important world issues seem to "drift". Is it any wonder then if people are losing faith or find it hard to deal with all the things happening around the globe?

Because the Great Shift through 2012 is an "invisible", global and highly complex phenomenon, it can only be mirrored in our news streams, people movements, social, economic and political systems. We won't hear about it from our established institutions or leaders that are often trying to cover up the truth behind the problems concerned as they are with the "status quo". So, the best way perhaps is to figure out what is really going on by listening (more) to the voice within? How often have we been told not to trust what we see, sense and feel inside even if or when it matters the most? Now, more than ever, we may have to think again ….

As we move through the Shift, more questions than answers may arise as we find ourselves in uncharted territory with regard to our future destinies. With or without a compass, the direction of the needle may change so to speak. Whatever authorities or institutions have to say, they may fail us if we only rely on them to figure out what is right or wrong, up or down. We have to become more resistant or persistent perhaps, as we strive to find answers to our challenges and solutions to our problems because nothing will be entirely the way it once was, again. It is pretty obvious that people who do not know this cannot help us understand what is going on. Rather they could become an obstruction instead of the assistance we might seek. And even if somebody truly knows, it is often not the truth that wins the public opinion. So, where do we turn to if we look for needed changes or necessary reforms?

The best bit about the years ahead, is that they will be very different from the past. As each year brings us further into the future, almost like a ravine, it could trigger more changes or experiences on top of an already big stack of world events. The winds of change are here and tell us about an invisible, significant force that is triggering a global change of events. Even if what we experience is initially only occurring behind the scenes or outside the normality of other peoples' frame of reference - we should not forget to listen to ourselves - trusting what we see, sense or feel is for real. After all, who understands what goes on in your life or world - better than you do?

According to the Mayan Elders the Shift will be experienced VERY differently by each of us, but we must not forget we are in this together while we are on an evolutionary time schedule. Only the outcome of the Shift through our shared soul experiences is blowing in the wind - not the wind of change itself. The universe at this time is bigger and stronger than any of us. No matter how we choose to look at it – we are at a intersection, crossing or choice point right now where we are given a choice - to either go with or against the Shift. There is no way we can compromise it completely. It is, however, up to each of us whether or how we will respond because we always have free will as human beings.

Nevertheless, some of this may be overruled if we are not careful and prepare ourselves accordingly for what is forthcoming. If we do - what will come out of our efforts? Could it be that we will discover that we are greater beings, with greater potential and possibilities than we have imagined, been told or ever thought we had? Can we prevent another war? Of course all of this "consciousness raising", human or planetary evolution stuff needs further investigation and exploration...

In a Sea of Uncertainty and Change

Whatever way the Shift unfolds within our thoughts, words, feelings and actions through our human interactions - we are ultimately responsible for each other and our shared evolution. The universal web of life interacts with us on a daily basis, whether we are aware of it or not, and has the power to affect us all at some point. Particularly a galactic alignment may trigger the feeling that - instead of surfing on a simple "solar ray" - a "crashing wave" may come suddenly or sweep us away. It may initially go undetectable. This is why it is important to stay more conscious or alert, by keeping eyes and senses open while trying to understand what is happening to us and occurring in the world.

It does not matter whether you are rich or poor, young or old, spiritually aware or too busy to even care. We all have to navigate through a sea of change or whatever is coming our way at some point in life. Anyone of us could feel something anytime, be it big or small. Could be from our kids running amok around the dinner table, to a big multinational corporation suddenly going bankrupt. As the Shift picks up speed and momentum, anybody with an open awareness can see that the course of events happens faster now than just 5-10-15 years ago and how much this is changing our perceptions about the world? This is just one of many impacts of the accelerated time and evolution, according to the Mayan Calendar and prophecy.

No matter how the impact of The Shift might be felt, from minor glitches when using electronic equipment, to a possible breakdown of telecommunications or our financial systems, social or political unrest, acts of war or terrorism or other unexpected events - the sea of change is here and its impact on our world.

As the Shift of the Ages unfolds, we can do whatever to adjust or re-adjust – but the old time or sun cycle will come to an end as a new one dawns. We are slowly, but surely beginning to understand that we cannot continue along the ways we have. The universe is giving us a much needed wake up call – in the 12th hour. We now know that global changes through human doing or undoing could affect our future food sources, economic, climate and other systems irreversibly. If the Sun suddenly erupts with solar flares our telecommunication satellites could be knocked out just as earth quakes that trigger tsunamis around the Earth's crust, could disrupt nuclear power stations and the livelihood of millions of people overnight. So, what are we waiting for?

A Wake-up Call?

If the Earth is going through "shakings" or we are going through "wake-ups" – (how) will our ways of lives change? Of course that depends on us. The Shift is a powerful shared experience or "cosmic triggered life-event" where few can stay forever completely immune with so much occurring on a global scale. Will you not agree?

What is being exposed in the outside world is equally reflected inside of us as some of the ways of life we have gotten used to become "unstuck" or "unglued" – or we are brought out of the usual "anchoring" - or "foothold". As the galactic alignment connects with the Earth's quantum field – we are also being stretched outside-in-and- inside-out.

Instead of trying to avoid it or recreate the past paradigm or evolutionary cycle, as we as humans may have a strong tendency to do out of a habitual feeling or need to stay "comfortable", we may have to step out of our comfort zone and open ourselves to the unseen or unknown. If we can do so as part of an individuation process we are given a golden opportunity to stretch our understanding or consciousness.

Next to expanding our minds, the Shift can open up our visions and ability to create something new out of the old. If there is anything we need to do as human beings it must be to evolve or find ways to evolve. Even if we have to consider some things we are not used to –or it looks like something outside of us is "asking" for a change, we need to look more closely, to see if it relates to a new Earth reality emerging.

A sequential series of pulsating waves from the galactic centre of universe contain increased energetic frequencies or signals of such magnitude, that it is "pushing us forward as a species to transform, change or evolve by allowing us to grow through a narrow window in time. This has not happened to us in this way before, so there is no rule of thumb on how to do it. Each of us holds a universal connection or key to the whole and because there are as billions of us, there are billions of ways to evolve or more as we can always go to a higher level from where we are now. Not two human life experiences are exactly the same at any given point in time, although the same energetic or evolutionary laws apply to each and every one of us because we are built of the same "universal substance" or "matter".

If we continue on a negative path of hate, an eye for an eye, destruction of nature, of fear and egotistical greed, however, it sends out "disturbing" signals and we could enter a "wrong" potential evolutionary time-line of future destruction or chaos. That is why we are equally important, if we consider our consciousness as a platform on or from which we can evolve, consciously, by realizing that we are all part of a greater organism or whole, and if we respect one another, are grateful for what we have, and take responsibility for our planet, then we could move directly into a positive future time line or a Golden Age for humanity and our planet. It is never too late to make a change.

The Sun and the Galaxy are awaiting our decision(s) – right now because time is running out of the old cycle. What we chose individually and collectively will shape our future destiny. There is no way around it. Whether we go through a time of suffering and destruction or we find ourselves united towards a positive purpose and peace – the decision is ours - we cannot rely on outer sources to fix the future and the world's problems for us. If we are to get across the Shift all in on piece – we need to get our act together for our next evolutionary stage or purpose, as soon as possible. Are you ready?

Personal Change(s) – Embracing the Unknown

As the Mayan Calendar tracks the evolution of consciousness, the Shift is an invitation to upgrade ours, to change contexts and learn to see things in (whole) new ways. This included getting in touch with our energetic (sig)natures (that some will call spirituality), and how our lives are created by our thoughts and emotional beliefs, which are basically the way we direct our personal energy.....towards something...

What is this about? In the end it is about a lot of things; such as understanding the spiritual beings we are, instead of believing we are simple (by)products of our history, family relationships, past memories or pains, our religious, educational, cultural and societal programs or conditionings etc.

We live in a world – particularly in the western societies that are still governed by the ego and the intellect obsessed as we are with externals, particularly where materialism, career, love or relationships are concerned. People often seek externals to feed themselves or rather to avoid emotional pain or confront the ego, in order to be noticed, move up the career ladder, look to others or seek to be famous and blah-blah.

There are few influences that teach us to investigate our inner world, which is really the only (spiritual) world, we always have at least from the soul's perspective. Rarely do we seek within to ask ourselves questions about what we are doing or why we are really here, including our values or true needs in life or relationships. Yet it is becoming clearer that now might be the time to slow down a little to pay attention to that part of yourself and ask some profound or life pondering questions. Only you know what needs addressing or questioning and the answers to these things that are really important to you.

The winds of change are invisibly affecting people all over the world in increasing numbers. People who used to be very outwardly focused, are now questioning their old beliefs or ways of life and are thinking about making small or major life changes that they would not have considered before. This is happening not just to people, but also to countries, businesses, organizations and situations that we would not expect it to happen to, normally. This can be both challenging and exciting, as the impending sense of change can make us feel increasingly uneasy, unsettled, uncertain or unsure about what to do while bringing up a round of old fears or issues that somehow need to be released.

A roller coaster ride of change often means steeper ascents and deeper descents with unexpected twists and turns in between the curves. As we as people become challenged on inner and outer levels through 2012 and beyond, wherever we are, this means, we might need to find new ways to use our personal energy or life strategies compared to the ways we have been used to using them before, as we go about our daily living. Moving adequately into the future may require new tools or skills to see us through on personal, collective and organizational levels. There is no guidebook or scroll that can tell us exactly how to manoeuvre the inner and outer curves of such a shifting ride.

To be able to do that or "get there" as whole human beings, we have to learn to work with our personal energy by moving our emotions and spirits in deeper ways in order to rise beyond our present ways and means in spite of whatever uncertainty.

Security means nothing, but complacency, really, to our present circumstances or reality. It is merely an ideal implanted in our minds to fool or trick us into believing that we are going to be secure or "safe" while fearing the unknown, that life in the end or at some point always confronts us with, but if we remain in the safety zone of the well- known, there can be little or no spiritual growth from an evolutionary point of view. When we embrace the unknown, we can take a quantum leap in opening the door to our part of the Divine Plan along the next corner of our soul's journey or the collective reality.

> *"The Shift is a surge of energy from the universe or cosmos - straight in (y)our direction."*
>
> *- Peter Christian*

Changes in the Big Sky

The universe is giving us a lot of signs & omens that something HUGE is likely to happen or happening. That something is going on in the Big Sky, already, is not an overstatement. It has been foreseen by many astrologers.

Astrology is the mother of science and mathematics and despite denials of empirical evidence, science does accept the basic tenet of astrology; that the positioning of planets affects life here on earth. Why else would science be concerned about the increased solar activity that might interfere with our high tech communication equipment and satellite technology? The Sun has been relatively quiet in recent years but is about to burst into a new era of radiation. Supposedly, around 2012 according to NASA. 2010-2012 happens to coincide with a number of unusual planetary alignments including squaring of the planets Jupiter, Saturn, Uranus and the new/old dwarf planet Pluto – through a cardinal T-square that is almost synonymous with 2012. At the heart of the matter is the square of Uranus and Pluto (almost conjunct Galactic Centre), which makes the 60s look like a walk (holding hands) in the park. The square is joined by Jupiter, Saturn and a number of eclipses. All of this is lined up with the Aries Point, magnifying or focusing the "new age" effect. Everyone is affected by this astrology no matter their sign, but some people experience this more than others.

Some people feel it as calm weather without needing to worry. There are some individuals and some charts aligned harmoniously with the current astrology process. Others are in for the storm of their life. When you look back on these years you will be amazed at what you experienced, but only you will know what it was and how it may have changed your life, who you became or what you accomplished. For many, there is quite a bit of unrest, turbulence, pressure and weirdness or they may be wondering what is next.

As we move through these rare astrological alignments during the Shift surrounding 2012, what we initially perceive as changes by coincidence, may be more outside or universally orchestrated than we think as part of pieces of something greater which relates to the bigger picture "out there". If change comes knocking on our door "down here" – it happens rarely when we only open it ourselves. We are connected to the outside world around us and everything starts somewhere. Nothing is always the way it appears on the surface. Even when life delivers with a "masking effect", the point is how do we relate to it? Particularly, when we are used to a slower moving evolutionary pace in the past?

If you have been living a part of your life in the most unconscious way, devoid of connection to the energy of the Big Spirit, tasting staler than three day old bread, then you need to open your mind to the idea of change. The universe does not like it when we sleep through our existence; or when we avoid getting in touch with our life's lessons. That is because our souls are on an evolutionary journey and we are supposed to grow in and from this life. If we are not feeling or embracing any need for change as we go along in life, how can we expect the universe to reward us with the wisdom from having new experiences?

Some day, the universe or something might come knocking on our door, waking us upside the head. Nobody ever said God had tact. If we are to see through life with our heads upside down, instead of spinning around, we need to prepare for whatever is coming our way in life. The X-factor of successfully dealing with change, also during the Shift, relates to our readiness. How ready, we feel on an everyday basis to change what needs to change in life and leave what we cannot be changed to other forces perhaps.

As human beings we are, and maybe always will be, creatures of habit. We like what we know and know what we like. Changes are supposed to happen, when we introduce them. They are not supposed to arrive without permission, or so we believe, anyway. The changes we are dealing with now, however, can appear from out of nowhere and throw us off course.

What if your enthusiasm for positive change is waning as you envision the dreadfulness of what might lie ahead? Do you feel prepared to take on the challenges and opportunities that are sure to come? Are you willing to go an extra mile or deeper within to access a subtle, more stable sense of yourself? Are you excited about travelling an unpaved road to find a haven or a Heaven perhaps? Are you able to step back from the world's fear-based ways that too often throws us off our centre? No matter what, you might have to figure it out.

To be enmeshed in a shifting process, or the lives we face today, can take its toll or breath away sometimes from even the most experienced change agent. It is not unlikely we can feel impacted on a number of levels. This includes our life conditions involving home, work, finances, human relationships or perhaps our physical health.

Other key elements include our emotional stability and repetitive thought patterns which determine the meaning we place on current events coloured as they are by our belief systems. Examples involve lack, failure and the power of authority figures or other issues we seem to have no control over or cannot access. If belief systems are catalyzed, our ability to successfully respond to crisis could become compromised.

Most likely, you will be challenged beyond what you think some of your limits are. A part of you intuitively knows that much of your current world or way of functioning is perhaps somewhat out of sync with who you really are or where you want to be heading. Life is full of limitations and there are many ways we can mirror that.

Virtually every sector of our old paradigm, parts of our society or reality that have been based on fear, greed or ego-based manipulations could become challenged. If structures are out-worn they might need radical reshaping over the coming years or decades. The reshaping or reconfigurations involved will be global, social and personal. It is not just our personal beliefs about these things, but also the fact that our eco systems are currently pushed to the limit that is causing us be under severe "environmental" pressures, particularly when the world's financial systems are knee-deep in crisis.

If we are "in tune" with what goes on, we can better "attune" ourselves to what needs to be done, so we have to trust the knowing that we feel lies within us. How else can we begin to create new life patterns, based on the beliefs or the ancient teachings of our world - the teachings of love, truth, respect, integrity, honour and peace?

No matter what happens our "destination" is not really "out there". Rather, it is about an inner journey, a shift from old patterns, ideologies of thought and beliefs, to new ideas that are more in balance with the Earth, the Universe and our Selves.

Finding Our Centre and Balance Points

The Mayan Calendars can teach us a lot about how to find a balance between our inner (spiritual) and outer lives on the level that we have chosen. One of the key elements to do this, particularly for Westerners, is to pay more attention to the inner you or the moment. Easterners, who are often more familiar with the spiritual side of life, may need to learn to focus on other things, such as how to manage time in relation to the outside world or physicality by taking affirmative action. In the West we have often mistakenly perceived spirituality as something separated from ourselves with no relevance to the outside physical world, whereas in the East – the mundane, physical aspects of daily life, have sometimes been given less importance at least in the past. With the right focus on both the outer and inner and dedication to these processes we can all move to a new evolutionary level where both the spiritual and physical aspects of human life can be honoured and revered. One of the ways to do that is to realign ourselves with the power of the moment.

Awakening to the Power of the Moment

A key element to the process of bridging the physical with the spiritual side of life and vice versa is to enter the now, because the moment is the place where the inner and outer aspects of our human lives interact.

Although it is humanly impossible to be 100 percent (positive) in every moment all the time, It becomes easier, if we become more aware of how we feel and behave through each moment. Understanding the dynamic which govern your moment, although fairly simple, is absolutely, vitally important. The very first thing one needs to do to master living with-in the moment, in spite of being infused in a third dimensional physical world, governed by linear time, and the material demands of daily life - is to be fully active in the moment while you are there. If we are constantly projecting ourselves into the future, or living in the past as a result of fear, inhibition or information contained within our personal "databanks", we are unable to embrace the moment and discover (how to work with) its magic.

To keep your self more in the moment you have to learn to live with an attitude of acceptance or gratitude. Out of mistake, we often tend to look at what goes on in the moment and think otherwise. By practicing taking out five minutes, perhaps every hour, to sense your inner self and surroundings while being grateful for who or where you are instead of focusing on what you do not have or have not become, you open yourself to the moment. Even if you live in a concrete jungle this is not impossible. If we cannot afford to regularly shut our eyes for a moment and give thanks for our life – no matter what it looks like, how can we feel grateful and graceful about being alive?

The more you do, the more you add to being able to feel grateful for being alive in the present moment, and sense what is really going on there, particularly within you. The parts of you that may have been projected into the future or trapped in the past can then meet back with you in that moment of time to recognize what is really important to you. It helps to realize both the spiritual and mundane things that are really meaningful and important to your life.

Instead of focusing on what you do not have, what you could have become or that which should have happened, instead focus in those minutes on the present moment, the only moment in time that you have right now, that is guaranteed to happen to you - right now. Do you see how it is only the moment that can bring you satisfaction?

As a matter of fact this second where everything is happening to you is the only now that keeps you alive and therefore the only thing you can say is truly real or happening. You do not know for sure, what will happen tomorrow or next in life and what happened an hour ago is not important, right now. That is why the moment is so important.

If you are alive, you are meant to be in this moment, right now, experiencing what it has to offer, and that is how you benefit from every moment of your living existence. If you are in a linear frame of mind with regards to the time you have or have left, you tend to believe that something is not the way it should be or should be happening. In this way we fall into the endless trap of thinking we have to move (on) to something better - somewhere else – when essentially there is nothing like that from a cyclical understanding of time, because otherwise how could we be where we are now, if the opposite were really the case? As a matter of fact, we are where we are because this is where we have the opportunity to evolve and learn at the time.

The energies that lie in the moment are always purposeful no matter where, who or what we are (doing) even if we cannot see it. Not a single second passes without (there being) an imprint of purpose. Because of the many dynamics that come at us all the time from all sides of time, it is impossible to discern which ones are just impacting upon you and which ones you can learn something from. If you are scattered all over the place, dangling somewhere in the future, miserable or lost somewhere in the past you are not present to the possibilities of the moment. The here and now is what is really important if we are to experience the practical magic of the moment.

You may ask, what has all this got to do with the Shift? Well, it might have everything to do with it, because the moment is even more condensed as time folds in on itself during the galactic alignment. In this way it has a more powerful impact on your life now than it did say 25 years ago. Will you not agree?

If the majority of people in our societies, however, are focusing on the past or what could go wrong in the future, while preparing for a war, strategizing against or secretly manipulating each other, motivated by greed or fear that are out of control – the dynamic that creates the future is defined or created by these limitations. What then is being focused on or given the most energy and attention? What is being nurtured and created in the moment that defines our future? I am sure you know the answer to this if you have been living through linear time for the duration of most of your lifetime.

Now you can understand why it is really important to BE in the moment, to acknowledge and recognize that which is positive and good or that which you are grateful for being able to do there. The more you practice this, the more the focus can change towards a more positive outcome or future.

If all of us followed real time dynamics, the world would be altered, almost instantly. The restrictive codes of the old time's dynamic would inevitably change because the matrix of that energy would then have been altered. This is the basis of the alchemy of the moment and the science of transmutation of time: The power to transforming that which is based on fear and lack into something golden in consciousness or opportunity. It is an opportunity we all have, right now. It is for everybody to embrace the ideas of developing more awareness and presence.

Developing More Presence

Aligning with the moment is all about presence because it always sits right there back with us. Even so, many would say as Descartes: I think, therefore I am, but of course this can never be "a priori" truth because of its limitations. Thinking is not the same as being or being present even though many of us have bought into the opposite idea.

When we talk about being present, it is important to know that there are levels or degrees of presence within the layers of the moment. It is as if presence exists on a vertical scale or axis. As you deepen into presence through the moment, you may cross a point on the vertical axis where time particularly during the Shift "disappears". It is as if time stands still. Now you are in the eternal realm. There is no time. Or all time. Life does not exist outside of the moment and you do not either. You have to embrace this idea, if you are to become more or completely absorbed into the Oneness of Creation. After all, you are creation and so is your life right now – right? All sense of separation is an illusion as it dissolves into eternity. It is a highly exalted or realized state of being. At this level of presence, thinking is nearly impossible as an individualized being. Any stand alone thought would take you out of the moment.

Revelations can arise if you try to connect with the past or the future, but the silence in between is radically different. It is like being without thinking. A sense of knowing arises out of the silence, but it is radically different from understanding what occurs only from within the mind. To experience this level of awareness in consciousness, even for a few moments can transform every aspect of your life, including of course your sense of self.

Although you don't have to be at this deep level of timeless presence all the time, as you ascend on the vertical axis of presence, time once again becomes available to you. You are not its slave, but can use your mind to participate in the world of linear time, at the same, while you no longer get lost there. It is about becoming one with the presence instead of being lost in (delusions of) the mind. You can think consciously, and when you are done thinking, you naturally return to presence. You are no longer a victim to those endless thoughts that keep dragging you to somewhere in your past or somewhere in your future in a vicious circle.

The beautiful thing is that when you can move up and down the scale of presence at will, one moment you might be choosing to consciously think, the next moment, you are at the deepest levels of eternal presence and silence perhaps. You are no longer imprisoned within the mind. You can move freely between the world of the mind and the eternal world of Now. Your memories, beliefs, ideas and opinions still exist within the mind, but you are no longer identified with or defined by any of it. You know that only this moment is the truth, and you are relaxed in the simple power of living in its present.

You can also speak and act from Presence. When you speak from Presence, there is no agenda that originates from thoughts only. You are not speaking from within the mind. You are speaking and expressing from the source of who you are, which is at the centre of your Being. You are present as you speak and can act from the art of Presence. Even Einstein's most brilliant scientific inspirations arose from that source and not from his mind. In creative and inspired moments, the mind is simply a clear instrument of expression.

Living in the Present Tense

Living in the present tense is a prerequisite or approach to do something with or about the energetic shifts that will culminate in the years around 2012. In whatever ways for instance, we are reliving emotional experiences from our past (that still construct the circumstances of our life), or whatever future fantasy cloud our head is in, we need to get present to what is happening here and live it more to successfully navigate the Shift. There is no scorecard, no judges on the sidelines waiting to hand you red cards or demerits. It is about your willingness to be aware of what is really happening – and what your body is telling you about what is going on in your life. Only with the willingness and tools to gain a present-tense perspective on living can life as it is appear in front of you, or people appear to offer you its special gifts that give you the opportunity to learn to get to know yourself and evolution better.

Living in the present tense means being aware of all of you, with openness to experience what is there and not what you think will or want to happen. An open mind and heart sees and feels the truth. It is about getting your body and its connecting to the inner you or vice versa on the same page: The ultimate goal of that is to become who you really are - what your inner self wants you to be like - so that you can look at your life, seriously.

> *"You need only change because you are finished with one experience and now you are moving to another experience."*
>
> *– Anonymous*

CHAPTER 5

REVIEWING SELF

Releasing the Conditioning of the Ego-Mind

Unlike what many of us think – we are not always free, but often subjected to conditionings or entrapments of the (ego)-mind that paradoxically or ironically often make us think we are "free" or secure, even when we are not. This is why many people often find it so hard to make a change.

When we can comprehend or have a context as to what is really happening and why we have certain reactions within the ego-mind, in most cases the energetic resistance to change stemming from mental or emotional dis-ease is released and much relief can be experienced or freedom achieved on many levels.

For instance, mental blockages are being released from layers of the mental body when an individual reads and participates with the material, that needs addressing or releasing, just as a release of emotional trauma through a spiritual reconnection allows the soul to move into oneness with itself without the blocked or debilitating emotions.

This has been continually observed by psychologist and therapist alike that when any level of the four body systems – physical, mental, emotional or spiritual - is experiencing a blockage (such as a mental or emotional pain, physical symptom or spiritual disconnect), the mind can participate with clearing it by having a context for what is happening. When the truth behind the context is made clear, the block often shifts by unravelling its cause, releasing its "grip" in that portion of the mind and body's cellular memory.

If you feel guided to recalibrate, adjust or clear something within you which is in or obstructing your life (path) - it is suggested to hold that intention before going to the spot of recalibrating, clearing or releasing. By being willing to go beyond the fear, emotional or mental blockages you can more easily release such obsolete patterns stemming from mental imprints, emotional problems or spiritual trauma buried inside the body's cellular memory during the Shift.

Although it is whole bodies of work to understand physical, psychological and spiritual healing processes, the more we can resonate with our Inner Source (truth, love and wholeness), the easier it becomes, as this vibration frees and liberates us.

Letting Go of Emotional Issues & Pain

We all know that we come from a family of human beings, and that the early events in our lives form our view of reality, shape our personalities and direct much or our (emotional) behaviour, most of the time without our even being aware of them or that we are repeating actions that were drummed into our heads long ago.

If life was/is sometimes painful, rejecting or suppressing emotional difficulties that we carry within us from a state of delusion or denial about these issues can do more harm than good in the long run because it "numbs" or "freezes" the emotions which in turn affects the physical body and mental aspect of consciousness at a later time as well.

Holding onto emotional pain keeps us hooked into emotional patterns of powerlessness, need or want. All emotional patterns unless they are based in truth will create problems because emotional reactions that stem from a lace of lack - hold the resonance of lack and desperation which attract more of that.

To resolve our most deeply buried emotional issues we have to deal with them rather than skip or ignore them by making the subconscious more conscious. This is essentially what psychological therapy is about. Although we do not necessarily need a psychotherapist to do something about this, it is important for our mental and emotional health that we become more aware and feel through the experience of life. It is an innate ability we all have.

When we are in our emotions, we can feel excited or overwhelmed by the accumulation of (painful) memories or imprints from the past. If we are residing there, we may not want to relive such stuff, but by not looking at or feeling it, we often remain on emotional autopilot or stuck on the dramas that prevent us from being ourselves particularly with others – even when they might be trying to love and understand us. In this way our emotions can stand in our way if we do not separate them from our true feelings.

Listening to our true feelings – or body feelings - is perhaps the most important tool to releasing emotions or emotional pain, as this allows us to resolve emotional issues by leaving them (in the past) where they belong. In this way we can return to a place of inner truth, emotional abundance and fulfilment that holds a different resonance and vibration which in return can create more of that. Although this is often easier said than done, the results are positive and many because it allows us to open our heart and experience grace and wholeness from within which can take us out of the vicious circles of an unaware ego.

This allows you to open the highest expression of who you are with further implications for releasing the personal ones that are limiting or debilitating.

Since none of us live in a bubble we daily interact with family, partners, colleagues and others on this planet we share. What influences our brothers or sisters, husband, wife, boss or co-worker impacts the whole web of life. Or as the saying goes; what goes around, comes around. This is why we have to learn to become more aware of the emotional dramas that we carry with us because they affect our relationships and other people as well.

We all know what it is like to go to work and someone there is having a bad hair day. Or come home from a hard day's work to find our partner is not having the best of times. We hear all the time how people's lives are affected by the random actions of others – be it intentionally or not. As negative emotions often cause vicious circles this in turn recreates the issues that are causing them.

Each of us have to do our share to sort through our interactions with others if we are to clear, release or heal something that affects the group or collective consciousness through a similar experience – if we chose to do it this can allow us to shift together to a higher level of consciousness in our human evolution during a time of remembrance - in consciousness.

*"One's mind, once stretched by a new idea,
never regains its original dimensions."
~ Oliver Wendell Holmes*

What Does It Look Like If Someone Is "Shifting"?

That is a good question because we can change or "shift" in many different ways on different physical, emotional, mental and spiritual levels while we interact with the world around us or the world interacts with us. Here are just some examples or possible characteristics of a "shifting" process:

- More emphasis on clearing out old emotional "garbage", ideas or beliefs is one thing. Just knowing that "stuff" you are hauling around with you is not helping or serving you and feeling a need to do something about it is a very good place to start.

- A need to be alone more, perhaps without knowing why. Feeling drawn to take in less stimulation via media of all kinds including talking endlessly on the phone with friends and family.

- A need to revamp or let go of long-standing personal and/or working relationships, without knowing why, or perhaps feeling a little empty or lonely because you seem to be the only one around you who is doing it.

- An increased interest in mystical, metaphysical or spiritual subjects, or remembering your interest in such matters from the past.

- Knowing or feeling that elements of your life need to change, but not being sure why or what you are to do with it all or what to change them to.

- Feeling restless, that something is missing or need to be explored or turned around in your life in new ways or on different terms to find a new direction.

- Experiencing dramatic life changing events, such as a sudden loss of a job, loved ones, or a health crisis - apparently caused or triggered by something outside of you.

- Fluctuating emotions or consciousness – from states of absolute bliss, excitement or exhilaration to fear, anxiety, panic or stress.

- Unusual experiences in consciousness or bodily symptoms that cannot be easily categorized or diagnosed - including erratic sleeping patterns, energetic stirrings or movements, unusual tiredness or fatigue, lack of motivation/inspiration or general feelings of disconnection, or even depression.

These are all general, but if some or several of them make sense it indicates an inner knowing that things are changing or (need to be).

If you or somebody near you is feeling that things need to change, by all means, let them change if necessary! For those of you who are not sure about the why or how of these changes, let them be for a while and let your intuition be the tool or guide to tell you how or get you there. One of the most important parts of living through the Shift is learning to accept our process (as spiritual beings) by trusting the gut instinct that each of us have.

If a shift or change happens in different ways or stages than you would like, by all means look up a health care professional or other person who can help you if necessary. Shifting happens according to our individual and group soul's stage of spiritual evolution, so whatever happens, happens as part of different ways of approaching the change, how we evolve, move through life and so forth.

If we can remember that the feelings of uncertainty are only temporary, and result from something quite amazing or truly holy occurring, it can sometimes make the process a bit easier to digest. Particularly when we stay focused on positive and happy things and in the company of loving and caring individuals, that support and understand us, these things will then be what are magnified, when new or shifting energies arrive and rev things up.

What happens if people near you don't want or refuse to understand, see or feel any shift, if you do or feel an urge to? Do not worry – not everybody does or does it in the same way at the same time. Maybe nothing or little will happen (in your relationships). Then it is okay as it is too. If people do not shift, or change in the way you thought, anticipated or hoped they would, there is no (universal) judgment. Free will is a birthright that comes with human existence – after all it is one of the greatest assets we have to learn from life's experience. The biggest thing is that if you feel drawn to let things change – let whatever needs to do so - by trusting yourself and your instincts. It is probably no coincidence that you have come across this material or other information about 2012 that for one reason or another resonates or is meant to guide you with your life's process or journey.

One of the major parts of "shifting" into a new (whole state of) consciousness or reality, is coming out of being in the fear-based ego-mind most of the time and into more self-trust and the (inner) acceptance that is crucial to evolving to find new and higher choices and loving yourself and others more. Fear is what keeps lots of people trapped or prevents them from opening to what is being written about here. According to the Mayan Underworlds (the previous cycles), part of the human deal the last few thousand years has mostly been figuring out how we could survive or be empowered while being told what to do from above or told we are not powerful by the establishment or people in power running things.

There are lots of sources about this, but the main thing to get here is the deep and serious level of spiritual evolution and health creation that results from understanding in what ways we are making limited choices for ourselves based in fear instead of making our choices or solutions from empowered new attitudes and beliefs of trust, love and truth with our spiritual power intact.

Moving Beyond Fear

Every situation we encounter contains the tools we need to either choose fear or to choose the path of evolution and growth. Fear works in two ways; it constricts and attracts everything that vibrates at its level and sends out a vibration that hides all options that vibrate higher than it does making them "invisible". That is why in our fear we are often unaware of any choice that will move us into a different choice or higher level of evolution through transformation or release.

Why do we as humans carry so many irrational fears? Where do they come from? How can we transform them and get rid of them? And why do they make us so powerless? For the sake of empowerment, loving ourselves and connecting with our authentic self, it is important to realize the grip holds of fear. The first step to let go of our fears, is to understand the nature of fear as a human emotion which is often irrational and un-justified – with no apparent relevance to the cause of what goes on in the outside world around us. Fear is something we are making up – it is not real. We simply feel fearful inside without knowing exactly why. When our fear is of such inner creation– it turns into an illusory monster that works against us, so that we are irrationally stopped in our tracks and keep repeating this pattern over and over. Perhaps even till the point where our fears are confirmed in the outside world again and again in a vicious circle as it connects to others.

If a certain fearful energy pattern keeps repeating itself within our consciousness – it is time stop and ask why it is happening or what is causing it. Few of us are told to do that – so we tend to skip over whatever fears - and let them restrict or control us instead. If we can get past being afraid of fear, fear can actually be a good indicator to reveal the very place we need to go to let issues go.

If you have irrational fears you need to be asking yourself some of the following questions:

What fear is being presented here? Why is it being triggered?

What are the other choices my fear is blocking?

What is the higher path that leads to a different choice?

What do I need to learn or know in this situation that will help release my fear?

This is by far better than projecting whatever fears or anger onto the world around you, where they might trigger your neighbours or vice versa - till fear goes around the neighbourhood. Only if we stop and notice what is really going on, can we start to see the forest behind the trees: Our fear is rarely a big issue, but a limitation or debilitating aspect of our ego consciousness. Together with our emotional mind, fears keep our consciousness in a state of not-being present or in control from where it cannot expand and progress our human evolution on life's pathway. When fears keep or maintain a grip hold on our consciousness and possibly self esteem, we cannot hear our own truth or feel good about being ourselves – so the first step is to listen to yourself without fear - feeling through the fear by doing it (what you are afraid of), anyhow.

By releasing fear we can learn to surrender to its opposite; that of love. It is essentially important because once this has been achieved, we are much better able to enjoy and live life, feeling and knowing who we truly are inside which is more likely to take us to places where we have been yearning for or wanting to go for a long time: A more loving, peace filled, abundant, healthy consciousness which can then reflect a similar life force and planet. Just think about how much energy is invested in fear, guilt, stress, worry, shame and the like (which binds us into thinking we are not ok) and how if each of us - would transform our fear into love and truth - the volume of our personal energy would be increased so that we can choose a new pathway of healing and liberation instead of remaining slaves to the fearful ego mind.

Every choice we make, to move beyond fear is a good one made with the knowledge and understanding we have at that moment. The choice is always ours it is a continuing journey. Often the choices not taken, however, are those that are hidden by our fears.

Once we release the fear, other things can be chosen and other paths taken. Then we can remember that we have unlimited choices and that what we might have chosen in the past could have mirrored our fears, doubts and misuse of power. Regretting our choices, however, is always a misjudgement that puts us in the past, instead of the present moment where we are ready to make other choices and become aware of them.

As we move through the 2012 alignment, it is a time for us to have courage and not be afraid. We are going to have to be clear headed and open-hearted in relation to what is next. What we need to do and when will then become available and the options and how to choose them more clear...

Trusting Our Intuition – Releasing Doubt and Uncertainty

How can we be sure that the next step is the right one to follow? We cannot, but once we have dealt with our fears or are dealing with them, it becomes a whole lot easier to read, trust and follow our intuition. Although we can never be 100 percent sure that we will be safe or what is going to happen next in life is ok – without our intuition, it is easier to get lost. If we live in a linear time mode or fashion where our mind wants to "know" in advance through changing times it can be unreliable and dangerous because the future is always in a state of flux. Trusting a move from point A-B in life in a strict linear fashion without consulting our intuition can turn out limited or dangerous because we have to allow for the unexpected in life, to strike a "right" or better path ahead. Learning to intuitive hunches can save us a lot of trouble in the long run because what leads from point A-B may encompass C which in turn may lead to a better point D. It is particularly important to use your intuition during the Shift where things may have a tendency to flip-flop between two realities or are "up in the air" so to speak by changing tides or waves before unpredictably "landing" somewhere.

We cannot be sure that a set course, direction or action - we or somebody else may have chosen - will actually turn out or end up the way it was intended or supposed to happen. We sometimes need to "wait it out" so that our intuition can give us the right information we need to make better, more aligned decisions and choices. When the intuition arrives it often comes at the right time like a flash of lightning telling us about this or that or how things will be(come). By trusting it rather than ignoring it – we can learn to trust in a higher source or power that can work out an even better outcome or path for us. In this way our intuition is a magnificent tool or a fascinating key to life which can turn what we think is going wrong upside down, teaching us something unexpected or how we can manifest a different outcome. It is really all about learning to be in sync with the RIGHT TIME, RIGHT PLACE.

Being In Sync – The Synchronicity of Life Events

Most of us have been taught or learned through the linear ways of thinking in the old cycle that if something does not work out or go the way we intended, there is either something wrong with us or we need to manipulate reality in some way – to get what we want, reach a particular goal or outcome before it is too late. In fact it is sometimes just the opposite. Our ego is not always in control or meant to get what it wants through a rational approach. If we can trust our intuition more and let life and the universe work for us in ways of the unexpected without filtering out hints, clues or signals we are receiving from our surroundings about the course of action we need to take, the universe can reveal the right thing or move when the timing is more right.

Although it can take a lot of trust to follow one's intuition or the voice of the universe in this way, the results of being in sync can be quite extraordinary when synchronicity points to the meaning of events in the personal or mini universe in relation to the scheme of the grander plan or universe. People, who find themselves in alignment with the inner and outer aspects of divinity in this way, may suddenly experience in real life what is called magic or miracles. E.g. you think of someone or something you need – and they or it suddenly appear at your doorstep - such as fresh apples out of nowhere for an apple pie you have been thinking of baking – or the right person for the right job you have been looking to get done. Synchronicity points in the direction of our interconnectedness with all things and by opening ourselves up to its force we can better experience or get the feeling of being in flow where everyday life becomes less of a "struggle". It is a very different approach that teaches us about how to stay in the flow of things and understand their meaning for the highest good of all concerned without losing the sense of purposefulness.

Stepping into FLOW

Talking about synchronicity, can/will we enter more of a synchronistic flow of life events during the Shift? Perhaps the question is opposite: If we are constantly in our heads - chattering with mindless thoughts – without listening to what we feel by following our guts, how can we step into alignment with a divine or synchronistic flow of events at all? It is about alignment with an inner "knowing" about how to connect with the right place, at the right time where our "inner" and "outer" world are more perfectly aligned. Only then can the "universe" seemingly from out of nowhere turn events in our "favour" or "direction" so that we experience the meaning behind "divine flow". By sometimes letting go or letting God can show us the way to a feeling of exhilaration about entering the right moment where we are suddenly becoming empowered by a force which is greater than our selves. Although this is clearly easier said than done, if we practice the art of going with the flow of life events in this way, we can perhaps better experience more "green lights" instead of banging our head against a brick wall.

Our Global Tipping Point

Until recently, reaching for a more spiritual point in consciousness has been a challenge for much of humanity. The stages of human spiritual development have ranged from basic "caveman" followed by spiritual stages somewhat similar to the ideas of human development proposed by the humanistic psychologist, Abraham Maslow where our basic human needs have to be met before we can attend to higher needs, such as self actualizing and spiritual ones. For instance, we would struggle to move to a higher social level where we can show love and affection without our basic physical needs for food and water being met first. The stages of transforming our lower or ego based self to a higher social or more spiritual level in consciousness, goes through the heart.

Coming to this understanding is where much of humanity right now finds itself along a crucial tipping point. We can either move up the evolutionary spiral or ladder, by embracing a more heart centred, spiritual consciousness based on love and unity, remain or clime down to the lower stages that belong to the old evolutionary cycle where the feeling of suffering or separation is more the norm. The wants, needs and demands of our survival based ego cannot be our primary objective. Although the methods of finding a more heart centred or "illuminated consciousness" has varied greatly from East to West, the eastern style has traditionally been a more direct approach, while the western one historically approached attuning with an outside divine source. Today, as we find ourselves in the middle of a Shift, humanity as a whole stands only a few steps from balancing Eastern and Western way of thinking through a more heart-centred consciousness. It is really about bridging the two brain hemispheres and hemispheres of the planet as well. Whether we are Chinese or American does not really matter.

When enough people join the "migration" to this New Reality in a time beyond 2012, a tipping point can/will be reached and the world will be consciously transformed. Every step we take to foster a more heart-centred consciousness particularly within our human relationships consequently affects the global "mind atmosphere", until it becomes fully "visible". Every time for instance, we remind a friend or a loved one, we share daily life with, of their spiritual nature, by showing our love, affection, appreciation and gratitude, we help the world move one little step closer towards the critical mass which can bring us to the next evolutionary stage. If this sounds like a utopia, what do you think the world would look like if we all did our share? According to the Mayan Calendar, the tipping point is almost here, so it has never been more urgent or easy, than it is now to actually do something about it.

Spirituality in the New Millennium

If man is more interested in spirituality or seeking God more fervently now, is it because the 2012 Shift of the Ages is bringing in more truth and light? In the old cycle the search or quest for God has been mostly associated with mainstream religions and faiths, but if we are looking for non-dogmatic teachings or teachers, we may have to leave some false beliefs or ideas from religion or about being spiritual behind. None of our worldly religions seem to truly answer the most important spiritual life questions, we are facing today. Although this is a highly controversial issue with so many people still adhering to a specific religion; through controlled transcription upon transcription of ancient spiritual texts they have managed turn metaphysical truths into religious dogma and rules about what to do or how to behave if we are to lead spiritual lives that align ourselves with God. This is not satisfactory. How can we grow spirituality through concepts of Hell, sin and guilt? Only spiritual beliefs that are truly anchored in our hearts can change the way we look at life and the ways we behave. At this time, it is about reshuffling what it really means to be spiritual.

When we realize that our inner core or true self is not separated from God – but a part of or an expression of the God Force, a spiritual path opens up that becomes an inner path or journey without too many religious instruments. It is a bit like practicing acts of God, like Jesus, Buddha and other spiritual master figures did, in a more meaningful, personal way. These figures originally took the inner path and were actively practicing it through essentially seeking spiritual human values such as love, compassion, as you sow shall you reap, good deeds, healing through God's grace and so on.

Instead of asking for outside religious guidance, what if we looked for answers to life's deepest or most important spiritual questions inside? It is not all about following certain religious beliefs, is it? It is about becoming one with an inner knowing of certain spiritual truths inside of us which we then can radiate outwards.

Spirituality is perhaps the most complex, universal subject, yet quite humble and simple in down-to-earth human terms if we allow ourselves to align with and practice it in every day human life. Everything is spiritual - meaning all is a part of Creation, God, the Cosmos, light or whatever you want to call the spiritual forces. The Great Shift will probably help us finally come to terms with what being spiritual or spirituality is REALLY all about.

PART 3

PASSAGE TO THE NEW EARTH

> "The best way to predict the future is to help create it."
>
> – Abraham Lincoln

Photo (previous page)

Apollo Earthrise

Image credit: NASA

CHAPTER 6

NAVIGATING THROUGH THE TRANSITION

Assimilating Radical Shifts Within and Without

Today, as humanity stands at the edge of the 2012 window, increasing numbers of people are realizing that we are alive in a time of many questions, incredible potential on the cusp of many changes. In other words, these times are not ordinary.

There is something that stands out about these years in which we are living: a calling to greater adventures, and the appeal of letting go of the limits and the fears that we used to define ourselves in the past. Although there is plenty of media focus on our changing world, it is so much more than that. If you are a bit sensitive, you might be able to feel the undercurrent of change even if you do not regularly watch the news.

Do you sometimes feel more uncertain about the future than you did a few years ago? Or are you excited about the many changes? As we observe our changing world's fragile state, we may wonder what the heck is going on, happening next or whether our humanity can come together for some much needed solutions for a growing list of global concerns. You certainly do not need to understand politics or economics to know that our planet might be at a tipping point. Uncertainty is all around us.

Similarly, you/we do not need to be on the path of conscious awakening to feel an excited acceleration about our time. Although everybody is working with the same 24 hours as you are, and many are feeling the stresses connected with the pulse of modern society, it is more than simply that or growing older feeling you have less time to live in. To many it is as if time is moving too fast and there just is not enough of it to do all the things that need or have to be done. One way or another, the way we experience time is not quite the way we used to. It can be because our perception of time is accelerating outside the linear time of the clock where time is becoming more elastic as it bends and stretches and contracts with the holographic or quantum factors of the Shift.

How do we make sense of the linear yardstick we humans have called time, you may ask? How can we intelligently approach the topics of time and the massive shifting occurring through 2012? What is the bigger picture that we need to understand here? How do you/we stay sane if the state of our world looks like it is in confusion or chaos? Can we become friends with these times of change – rather than crushing under the changing (time) waves?

Below are some ideas, hints and suggestions that relate to this and how we could resolve some of the collective and individual unsettledness of our process as we go through the Great Shift of the 2012 time transition.

If we feel a stirring in time we can ask ourselves if what we are feeling inside is related to pressure coming from the outside. In this way we can better figure out how to deal with individual occurrences in relation to the collective time shift of 2012. By asking ourselves certain questions we can become more aware of what is going on when or where and figure out how to deal with it all.

Even if we feel "blank" or as if nothing particular is happening, it might only look or seem that way on a surface level for a time. As times, people and places are shifting in many different ways during the Shift of the Ages, things that have stayed the same for a long time, may suddenly change next week, slightly more next month or even more dramatically in the year to come.

Make no mistake about it... We are not in Kansas anymore and there is no turning back to exactly the way things were or where we come from in relation to where we currently find ourselves... We have to move forward... The challenge is to do so in a smart or "enlightened" way... by being "Divine Inside" instead of being or feeling like "The Walking Dead".

No matter how we chose to look at the "End of the World" or "End Times" the fact is that the strong (evolutionary) winds of change that are up in the air can be quite disturbing or unsettling. We even see some of these reflected in irregular weather patterns; one day the sun might be shining and the next minute the sky disappears behind a giant carpet of grey clouds before it is pouring down with rain or we are caught up by a tornado or giant hailstorm. Haven't you noticed? Nothing seems totally reliable, predictable or the way it once was anymore in regards to the weather – or is it just the way we think or talk about it?

So, how can we be prepared? What can we do if are sometimes stressed, confused, tired or worried about it?

Could it be that we are in the process of dismantling an old or outdated future weather forecast in order to move forward into a new or different weather pattern which is not quite there or here yet? Could it be that we are in this transition zone either on an individual or collective level or both because something entirely new under the skies is emerging through our time? If we are neither an invisible nor a separate part of the whole – what we are experiencing or beginning to understand is perhaps how we are intricate, delicate pieces of it all and that no change, be it individual, global or universal, stands alone.

With something as vast as the Shift of the Ages – change can be associated with turmoil, confusion, difficulties, upheavals and challenges, but it could also be giving us many opportunities to discover new exciting possibilities if we are ready to embrace what lies ahead on that journey. If we can understand more of what is going on in the world in relation to what is happening to us, is it not likely that we can make something different or better of it?

Times of transition or paradigm changes involve both endings and new beginnings. It is a transitional time or zone where we learn to disengage from a past evolutionary cycle as we engage with a future, new one. Like any potential, dramatic change – it is not full of the type of conditions that we have gotten so used to, know or think we are looking for. Instead, it can be full of uncertainty, chaotic or shocking events that suddenly pull the rug out from underneath us like a tsunami, for instance. It can be a time which initially feels extremely unsettling. Just take look at the past few years; each month has brought new events to the world: Many are in the form of natural disasters, while others are politically or economically motivated. Have the past few years affected or changed you or other people's lives near you in some way? What has it done to you?

It is important that we do not get caught up in preconceived notions or ideas about (optimum) conditions of how things should be or develop. During such an unusual, transitional process, things might not be or turn out the way we think or expect. Instead the future is in a state of flux where we need to develop the flex-ability to make the most of who or where we are or what we already have to navigate through whatever choppy waters, swirling change or undefined territories that often precedes a new birth following the collective unsettledness...

The Collective Unsettledness of the Transition Zone

The Winds of Change are affecting the collective because humanity moves from one evolutionary cycle or era to the next. We have already talked about that. This includes increasing numbers of people(s) around the world who are now questioning old beliefs or ways of life while possibly thinking about making a change (of awareness) that they would not otherwise have considered before. This is happening to people situations, organisations and even businesses from all walks of life that we normally would not expect it to happen to.

The important thing to understand is that the impending (sense of) massive change can bring all kinds of issues or turmoil to the fore that make us feel increasingly unsettled as it brings up another round of what has been "hidden" which need to be released or old fears that apparently were not there before.

If turmoil, shake-ups or wake up calls occur, we need to understand that they could be more than simple, coincidental incidents and consider that they might be part of a greater chain of changing events. Things could happen all of a sudden because we are constantly moving, meaning if we continue to live in the past we are not going to align with the new space or time cycle which is arriving for our planet and humanity.

It embodies a whole new set of coordinates. Like a super tanker which has been set on a course, it takes time to find the new destination with a new set of coordinates. We cannot turn it around in the middle of the course when there are highly changing waves, we need to wait for the currents to settle a little or adjust its steering wheel in order to cross the seas and arrive safely in a port.

No matter how disturbing any lack of fair weather may seem, we have to find a way to get ourselves across or under the bridge that will take us more safely to the new haven. It encompasses our ability to join forces or work together using our ability to love and respect each other and resolve whatever conflicts or problems that might occur in the process. If things look like they are in strife or chaos for some time, it is because we get carried away by fears or insanity and forget to work together as a human team. Just like a sea captain who needs his seamen, so must the land crew which directs the ship find together and find each other to assist. We have to find our centre to stay balanced and avoid sea sickness no matter what, because we cannot rely on the outside world to "fix" this part for us.

If we do not pay proper attention to what the universe, Great Spirit, cosmos or whatever you want to call it, is telling us, it will throw a cog or two into the (global) machinery or wheel. We are already seeing that we need to make some necessary adjustment(s), otherwise we are moving towards a Spiritual Course Correction. If on the other hand we are willing and able to change whatever needs to change, we could see all kinds of positive things happening with our world that could take us another 1000 years into the future - safely. Either way, it is the beginning of something new. Not one (over)ruled by our human egos or ignorant corporations or governments, but one which allows the (Divine) Love of the Big Spirit to begin to rule the planet and its people. Does it sound otherworldly or too farfetched? Well, this spiritual transition or correction is going to be the way Spirit works because it is not ruled by our human will alone. Mankind was never given the authority by God or the Great Spirit to rule over others or it-self.

Every time it has been tried, it has failed miserably...turning out very wrong... leading to major wars or suffering. To avoid it happening (again) we really need to understand that the Great Spirit or Divine we all belong to is Love. If modern life has turned its back on the Love of the Big Spirit, know that Spirit is returning now and it will be turning its back upon mankind - unless of course we chose otherwise. Are you tracking on this?

We can either choose to work with it or disengage from the process (all together). Come floods, ravines or stirrings on a vast, beautiful lake; Spirit has arrived to take us with it in its wake. There is no way we can stop the movement of the universe...

No matter what happens to us "down here", there is no need to panic because such a reaction on a collective level would only be of an irrational fear based emotional kind where our egos are concerned and could make things spin out of control rather than allow spirit to resolve the situation(s) for us. As each of us are intricately linked to or a part of the whole, we need to find ways to deal with the collective unsettledness by focusing on the good WE WANT to happen – rather than the things WE DO NOT WANT or are afraid will happen. As long as we deal with whatever might come to the surface in a sensible, appropriate or adequate manner, there is no need to get caught up in (outside) turmoil or unsettledness. This would only add to whatever problems by giving more focus and energy to that.

Like a ship that is temporarily lost at sea or a small boat in a sudden waterslide we might lose our "grip" or control of things for a while during this Shift, but if we can just accept it by going with the process is it not easier to "change with the tides"?

There is no need to or reason why we should "go down" if we pay attention to the weather forecast. It all depends on the ways we choose to prepare ourselves for the weather or storm that is coming. Panic, shock or turmoil occurs only if we remain totally ignorant. It is so much easier to adjust and adapt if we are ready and prepared.

The cosmic wave or surge of energy that will be engulfing the planet around 2012, like a "big blast" can either throw us off balance or help us rebalance depending on our individual positions on board the ship. Whatever is being washed away from the deck perhaps needs to be cleared off the ship in the waves' wake.

This is only to give an idea of how the world is under pressure to change and transform its course – like the giant ship it is manifesting through the many small or major crises occurring.

This includes social, religious, educational, political situations and institutions that we may have relied on or taken for granted in the past.

It is more important now than ever that we do not fearfully cling on to old structures, outdated ways or habits because they can mislead us. Instead, we need to find the necessary courage, confidence and humility that will enable us to walk forward and create or solidify a new course of action in reality; based (more) on love, concern, sustainability and compassion for us all.

Anything that needs to be recalibrated in our collective mindset to become realigned for that new reality is up for "grabs". People who consider this astonishing or even threatening need only look back to see some of what has already happened as we passed through the previous Mayan Underworlds. Do you really think evolution can be stopped? Just look at how much has already changed in our human history. The further we progressed through these underworlds - the more evolutionary mechanisms or issues have been brought to the fore, so that we were better equipped for what was next. Although the present may look like it repeats the past – it never quite happens in the same way.

The years around 2012 works like a giant intersection. Not unlike two Mayan Calendar wheels that click into position to set a greater wheel in motion, we are starting a whole new time. If we think we need a "time mechanic" to fix some of the "squeaks" within the wheels, perhaps it is really more about how we fit in the holes or gaps between its rivets of (r)evolution before whatever "cracks" in time get out of proportion.

The Mayan prophecies say that we have an unprecedented opportunity to pass through the turning wheels of time into a whole new era of our choosing. As the final rivets at the end of the Mayan Long Count Calendar click into motion we are standing at a choice or change point in our understanding about the shared responsibility we have with regard to our planet as we pass this threshold.

The next evolutionary stage or level of our journey is awaiting us around the next corner. We can either seize the opportunity to chose rightly or make things right to finally live in global peace and harmony for the benefit of us all - or?

Destroying the Apocalyptic Myths (of 2012)

Although both modern astrologers and the Mayan Elders agree, there is a lot of intensity going on in the Big Sky during the years around 2012, let us take our eyes of the sky for a moment and look at how we can work through this time to destroy the apocalyptic myth which has also been created.

No matter how we look at it, the powers of the Shift are challenging. The 2012 phenomenon is putting pressure on us. Like travelling on a galactic spiral arm into the galactic centre's black hole of "gravity", we need to find a way to move through the eye of the storm one step at a time.

If we experience pressure there is always a way to find some "vent" – which can ease off the (outside) pressure even when we feel stretched or uneasy.

Although "the 2012" does not come with a recipe or manual it nevertheless comes with some important clues; we need to understand. We are in some form of transition, transformation or (re)creation of our personal or collective reality which of course relates to our future. The question perhaps is not if, but how we are going to get to the "other side" of the tornado's "spiral". Are we going to be broken or arrive in one piece?

The experiences and solutions will probably be as diverse as we are people on the planet. However, there is one common denominator; namely that we each have a human consciousness. We need to understand how our individual consciousnesses (can) work together instead of seeing ourselves as "lost souls" drifting or separate units in the grand wheel of life. If we don't get this, we are running a serious risk of splitting ourselves apart during the shifting process. Some will be for and some against it, but perhaps what we need to do is entrust the 2012 transition or "buffer" zone with more LOVE.

Unlike the rest of our human organs, we can always feel the heart because it is pumping life sustaining blood through our veins. If the heart is full of love it is more likely to keep pumping. Anger or fear seems to clog up the vessels or circulation. All members of humanity know this and have a heart and if we can focus on keeping our veins flowing with love rather than its opposite, isn't it more likely that we are going to feel less fearful or alone? It is both an individual and collective decision of course.

So, instead of asking – will we be safe or will there be a future Armageddon during the years surrounding 2012, perhaps you/we not need to ask how you/we can live in a way that contribute to the hologram of our planet and collective with ways that make us feel better, more caring and loving in relation to humanity to provide a saner and safer, loving future for us all. Do you agree? If you are thinking in this way, you are probably one person who is making a difference.

The Individual Unsettledness of the Transition Zone

If you are connected to and conscious of the cosmic waves or energies streaming into the planet - you may be feeling this already – but even if you are not – you can possibly expect a change period of unsettledness surrounding it .This is not to alarm or scare you....whatever you may feel, will have a tendency to come and go with the waves of the Shift as it slowly wanes in the years beyond the 2012 end date. It is important whatever we may feel because it tells us something about what is happening to us. None of us are immune to emotional changes or can stay totally unaffected by a "big push" of life changing events. Whatever 2012 and beyond entails for us personally, it has the momentum and power to change us outside-in and inside-out. In this way it has multiple implications – whether we are consciously aware - see ourselves as part of a change process - participating in a Shift or not.

Our ability to keep up with the quickening pace of life depends upon our ability to handle abrupt unexpectedness in a way, while we take care of life's many practical details. Even if we do not take any of this personally, we have to be conscious of our actions and always remember that our responses in personal life ripple out and impact the field around us as well.

We have to understand the act of compassion for ourselves as well as others because individual out-bursts can instantly change things in or out of form within the collective consciousness' matrix. For this sake, we have to focus on transmuting unconscious guilt, past "debts", persecution, religious judgments, the seeds of revenge or violence in our human behaviours based on dualistic, mental distortions and distinctions between good and evil as part of our history. By transmitting forgiveness and radiating courage from within ourselves to everyone we jump into the fires of love that eventually transmute and transform everything through grace. Each act of love and every experience of grace and peace moulds and moves the incoming energy into the highest potential for every living being participating in this sacred Earth event.

The cosmic trigger surrounding the galactic alignment will be a defining and most critical directional point for us and the planet in the years to come. How we respond to the event, through individual choices and action, casts the anchor for how we transition into (desired) new realities.

Would you like everything to stay the same? Most people would probably say no, but how much are you willing to see or experience change?

The individual unsettledness we may feel during the transition could be "triggered" from the outside, while felt on the inside – meaning the "inner life" or inside of your consciousness is subjected to outside spiritual influences at this time. Although most of us tend to think that we are separate from the outside world we all know how often it determines the way we feel on a daily basis, even though we like to think of ourselves as free-willed or strong human beings. It happens because the inner and outer lives we live are intertwined.

If you sometimes feel a sudden stir, change or moving energy within you, then you might somehow be more actively engaged or participating in the Shift.

It is only natural then if you feel something is REALLY going on within you and you are not making it up in spite of what the outside world might tell you…. The galactic alignment works somewhat like a vortex ring which can stir up all kinds of "dust" and moves our personal energy – the energy that you use to live and experience your life.

This can be felt in many different ways depending on (y)our sensitivity. For some it may feel a bit like a blow or tumble drier with hot air or gas blowing through or even around you in a similar fashion or motion as that of a vortex ring. You do not have to open yourself to it. It happens naturally and if you are a sensitive one, it may even feel like a "pressure cooker" at times.

Whatever happens can "bring things to the surface" or turn them around and sometimes upside down within or around you. The winds of change have no particular intention other than bring in more love and light, so it is really an opportunity to learn something about that on your pathway.

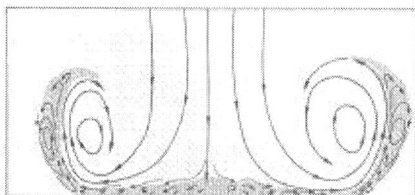
Vortex, microburst ring

If you find yourself in the midst of a "hot vortex" or "inner storm", somehow, knowing what it is, why it is happening and that it will not last forever, can greatly help and enable you to better deal with whatever is going on. Although it can be quite puzzling, confusing or even bewildering sometimes to be in a spiritual change process, it can also be exciting or exhilarating. Depending on how we chose to look at it, it is probably both at the same time. At least you know that you are not "losing it". As a matter of fact you are probably becoming more aware of who you are and what goes on in your life. With greater awareness you can better "ride through whatever storms in life" and learn or see what you need in order to evolve and progress. There is nothing wrong with that – on the contrary! Being fully alive means consciously getting to know your Self.

If you are transforming something in some way, what is happening is that it changes aspects of you and your life that may have become stale, stuck or dried up before they become outdated and need to be turned around to allow for something new, better to come. If you are courageous enough to see this as possibility, you can move through it advantageously to become a truer, wiser, more evolved or powerful version of yourself: You are the creator of your life – remember? If you are not happy with some of it – you can change you and turn it into a different thing. It depends on your circumstances as well of course.

We cannot always control what goes on in life, but whether we pass through it willingly or unwillingly – surrendering to the process with courage and trust - alleviates some of the difficulties associated with the ego's resistance.

Symptoms of change during the transition zone of the Shift, that we are going through, are not going to go away miraculously - when we get passed 2012, because we then live in a future reality, new Earth, better time, paradigm or era if we are lucky enough, thereafter, but if we can accept that we are in a process or about to evolve into something we do not know exactly what is yet, it eventually helps us express ourselves in a different way or to create a different reality, all together.

As there are so many of us on the planet travelling along different trajectories, it can be quite challenging individually participating in a collective shift or change process with all kinds of agendas going on at the same time. It can be difficult to see the bottom lines of the transition for you and everyone else. To figure it all out can be mind boggling, and most likely we cannot because there is and will be too much "stuff" going on in our individual and collective pools. Is it any wonder, if we sometimes think the world is drifting or perhaps coming to an end?

If we have lost our way or feel like we are suddenly unable to continue along the pathways we have been used to (in the past, old cycle or paradigm), it is because we are bound to learn to be alive or live differently in ways that will resonate better with our future reality.

Each of us have our own individual or personal reality, so, whatever change processes are occurring in the outside world reality may not be the same as the ones happening for you. There is even a tendency that the past, present and future are interacting or being reshaped in certain ways towards a new outcome as time stretches and bends along a curved trajectory. In other words, the future is not set in stone. Yet.

It is often difficult for our human nature to deal with such uncertainty, because each of us would like to feel "safe" as individuals while we participate in the greater context. Without a manual or guide which lead in a certain or particular direction we can feel lost. However, if our personal reality is changing and the world is similarly adjusting to change through its apparently non-cohesive yet connected parts, the more we can allow whatever needs to happen to happen, the easier it becomes for a new reality or paradigm during the collective Shift to emerge.

The unsettledness of this transition can kick up a lot of stuff inside and outside of us as we are consciously and subconsciously "working this thing out". We have to remember that whatever is stirred and moving is there for a reason. It can teach us something about how we can transform old parts or aspects of ourselves in order to move "upward" or "forward" on our evolutionary journey as (a collective body of) individuals.

We will need to embrace and work with this one way or another to strike a new way of being that elevates us to a higher ground. This does not happen overnight and takes time to settle of course as we gradually shift time cycles.

As we gradually step out of the old time and into a new time, it will become more evident that we are stepping out of certain old experiences, aspects of self or ways of life. It means that we might have less time than we think we have before we have to move to the next stage. In this way the Shift can be similar to a 100-400 meter hurdle race.

As we move from who we were or how things used to be into someone or something new, we can discover hidden abilities that we have that allow us pass these hurdles if we lift ourselves up and move forward. Instead of skipping the hurdles, we need to strengthen our "spiritual muscles" to jump over them which of course will be reflected by the finish line.

In this way the Shift teaches us how we can develop inner abilities as well as outer capabilities in relation to what we are going through. On the other side we will be stronger and better equipped for what is next. We cannot always see the road or lane ahead because it is "curved", but we can see that each moment is a step we need to take to prevent falling off by the way-side. It is all about staying in good flexible shape in the curves of a changing race within a shifting world.

It is about psychological management. We may have little or no control over what is happening "out there" in our life; but we do have something to say with regard to how we choose to deal with it. If we are not looking for change at least we can change our stance and not get thrown off balance by change. We also cannot allow potential shocking events to take away our focus or joy of life. In worst case scenarios we still need to find what really matters, has importance and is deeply meaningful to us.

If we find ourselves midway between panic and joy, it is better that we do not get wrapped up in whatever turmoil by letting ourselves be pulled into fear or too much excitement about this or that. We can stay expanded, real and authentic no matter what, if we trust our ability to know and handle whatever situation by opening and allowing ourselves to discover what we need to learn or understand from it to move forward.

While it might be true that we as human are always seeking new or better opportunities in life, our present conditions are always there to teach us about how to deal with reality. Restrictions are as much a part of being here as are opportunities. Even if they sometimes block our vision, so that we cannot see that a totally different set of more optimum conditions are actually here. During the Shift, these are optimum conditions for stripping off illusion, deep internal transformation, quantum breakthroughs, for getting real and finding our inner security or serenity that allow us to merge with the fullness of our true human being. Does it sound like too much?

An essential component of this navigating a new wave or paradigm is the ability to do things differently by using what we already have in life to recreate the things or conditions that allow us to go in new directions. In the new paradigm, we have to learn to be able to function with more fluidity, flexibility with love that can take off some of life's punches.

The rapids of inexhaustible change do not allow procrastination. The time for complacency is over because it is not going to take us to the bridge that crosses the river. We need to allow ourselves to "know more" rather than "know better".

Interestingly, it allows the more unlimited versions of our selves to come forward. It is needed because what looks "safe" and ok today may become useless or dangerously in our own way in the future because it might not take us to where we are supposed to go. The longer we wait – the more time we waste and will have to wait for what we really want to happen. Isn't it better to walk away from something or someone that we no longer feel is "serving" us than to postpone an inevitable finality or resolution? We need to become clear in order to stay clear about where we are going and adjust accordingly in each situation, so that we can solidly move forward with more self confidence, ability, integrity and trust.

For those who are spiritually awake, attuned or inclined this crescendo of energy offers the opportunity to receive renewed inspiration and highly creative solutions for future life situations. There are challenges up ahead for sure, but we have to remember that every challenge we face in life is also an opportunity that can teach us something about how to go beyond with greater stamina or vision. Above all, we have to remember during challenging times that our (cap)abilities are often greater than we think, but before we can do it we may have to face some emotional dragons...

Setting a Limit to Sit With Negative Emotions

Are you sometimes tired of negativity in life? Emotions colour our lives for good and bad with varying palettes. Sometimes we feel a strong emotional reaction to something that has happened, but emotions also visit us seemingly out of the blue, flooding us unexpectedly with joy or grief or melancholy. Like the weather, positive and negative emotions come and go, influencing our mental state with their particular vibration. Sometimes a difficult emotion hangs around longer than we would like, and we begin to wonder when it will release its hold on us. This is often true of grief stemming from loss, for example, or lingering anger over a past event.

Usually, if we allow ourselves to feel our emotions fully when they come up, they recede naturally, giving way to another and another. When a situation or emotion haunts us, it is often because we are afraid of really feeling it. Emotions like despair and rage are as powerful as excitement and exhilaration, so it is only natural to want to hold on to them or keep. Particularly fear. Certainly, we do not want them to over take us so that we say or do things we later regret. When we are facing this kind of situation, it can be helpful to ask, "How long do I need to sit with these emotions or feel blocked by them before they can pass? How can I release them to transcend my emotional personality? If you ask sincerely and wait, an answer will come. If you will, find a time to limit your engagement with that difficult emotion, sit down and make yourself available to the emotion that has been nagging you. All you have to do is feel it. Avoid getting attached to it or rejecting it. Simply let it ebb and flow within you. Emotions come and go, so you can trust that just as one reaches its apex, it will pass. Each time you sit with its presence without either repressing or acting out, you will find that the difficult emotion was the catalyst for much needed personal transformation and an important way to heal your life.

Surrendering, Releasing or Letting Go?

If any aspect of life or self seems blocked there is often a hidden reason for it that we need to uncover instead of holding onto the problem. No matter how difficult it may seem there is almost always a different way or solution that can take us out of a dark or blocked pathway. Although it can be challenging to surrender, release or let go of what we hold on to, perhaps it is for the better if we let it go. If it happens to be that way it can be because your soul is crying or calling out to you so that you can unblock the stuck aspects of your self.

Getting (emotionally) unstuck helps get out of old ways of life that are "fixed" and have stayed in the same rut for a long time. For instance - if we are "being asked" to abandon certain ideas of life or aspects of self - such as that we need money to be successful or to have other people around us all the time to feel happy, that are not aligned with our true Self at the core of our best interest, we need to open up before new ways can allow us to move towards contentment, joy or happiness in life. The idea that individual wealth, power or materialism must be pursued at all costs or that we have to live in a strict linear fashion belongs to the rational-material logic or paradigm of the past cycle. Although there is nothing wrong with money in itself, there is something wrong if we get trapped in it and cannot find joy in the simple or small things in life. A new cycle is upon us, so it is important to find the hidden gifts or gems of life by understanding the possibilities of interconnectedness, love, creativity and community that we can experience simply by being who we are and sharing that with others who understand us. Although this may seem a bit "alternative" at first glance, it might eventually become amongst our highest priorities, because it allows all of us to experience the abundance of life. Self-education or reflection is the key here to make a transition to live life more purposefully and meaningfully. We cannot take our old self or ways for granted if we are to experience a different state of consciousness or happiness that allows us to step more fully into our Selves instead of waiting for someone or something outside to come and "fix" life, this or that for us. We need to do what needs to be done to step into a new role or persona, by taking responsibility for who and what we are and how we feel about it. In this way we can become the "new change agents" that activate a new way of life with others.

In the Hall of Mirrors

Moving away from negativity and limitation in relationships involves letting go of the dramas of the ego. This is essentially about being more of who we truly are, by trusting how we feel inside the core of our being– instead of allowing fear, domination and other unconscious mechanisms and tactics to take over our relationships. People outside of us can only mirror something back to us that is already inside of us. For instance, that we need to learn to disengage from whatever dramas that are telling us or them when we are ok or not, when what we should really do is to simply be who we are. If we find ourselves "trapped" in private or professional relationships it is often because the ego gets in the way with its limited, self centred mindset that does not allow the truth or potential of a relationship to come forth. If we are not aware of these relationship dynamics or perspectives and defy other people by not coming from a rightfully justified place of love and truth, we trap them or they trap us in the lower aspects of the relationship where one half is controlling the other and the other is being controlled. So, how do we take ourselves out of these ego-based relationships and move towards greater self-empowerment, authenticity and integrity between ourselves and others?

One of the keys is feeling the energetic chemistry of what we are sensing and feeling. We also need to work with our response-ability – meaning ability to respond to the relationship in a responsible manner which neither leaves the inner core of our Selves or others out. If we are with the wrong person so be it, but if we are in the wrong situation with the right person, we need to think again about how we can resolve the situation through our personal actions and reactions. We always need to look at ourselves first in the relationship's Hall of Mirrors before we can make judgments about another who basically only mirrors aspects of our personality back to us. Once we do this it can have great impact on the way we are with others. How other people react towards us always teaches us something about ourselves; and vice versa of course.

If we can keep this in mind and be mindful about the way we think or act the whole issue of whether we or the other person is right about this or that becomes irrelevant. We will be naturally guided to say or do the right thing in the right context because our relationships will be based on what we know inside that can empower ourselves and other people through more conscious acts.

Facing and Finding the Authentic Self

If we are bound to face more of our authentic self during the Shift, we are not just resolving individual, but also the collective karma that relate to the ego-persona. Becoming more authentic means becoming more of (the real) you by stripping away layers of yourself that are really not you. It is about aligning with that part of you that is illuminating and full of radiating energy. There is absolutely no reason why you should be somebody that you are not, which in fact is the truth for all of us. When we understand that our trueness lies at the core of who we were, the day we entered this world as an expression of the Divine, even if we have steered away from this path and it is something that we have forgotten – we can rediscover and remember this truth.

In the old time cycle or paradigm many of us have bought into a wrong idea that it is better to suppress our true or authentic self for survival reasons, out of fear, because it might not be understood, accepted, loved, heard or whatever by others, but in the long run this does not work. Although we may have taken this for granted it is wrong, because we cannot suppress who we are forever. As a matter of fact, it kind of works the other way around. People can only respect and love us for who we are, if we stay true to who we are without insane, pleasing role plays, even if that means other people might hear or see something they would sometimes not prefer.

Our authentic self is the "hidden gem" or treasure chest that delves infinitely within each and everyone one of us. It is our uniqueness – that special part or place of us that is our most precious gift and which never leaves us. It is important not just to ourselves, but to other people as well and the world we live in that we let it out of the closet. We serve no one by hiding our light under a bushel. So, why all this cover up, if being true to who you are, is what also truly serves others?

This new way of interacting from our essence – instead of being right or dominating with or egos – relates to the new Earth paradigm in a big way because it works through acts of resonance or transmittance of truth, love and wholeness rather than their opposites.

Resolving Power Issues

Similarly, power issues have often dominated in the ways of the old cycle where our egos wanted to be in charge or take control of another in the world. In the new reality we have to become more authentically powerful; leading by example from who we truly are. This is our real power which cannot be denied or rejected. It can be reflected in the outside world through truthful interactions that are more honourable and respectful because they come from a place of integrity that works for the real benefit of other people and therefore our selves in the long run. As we pass through the Shift we are being asked to resolve some of these power issues by learning to apply a new way of empowerment to many of life's situations.

Anything that does not serve other people or the whole is no longer sustainable and will eventually be rendered useless in the new cycle that is emerging, because it is more about "us" than "I" – whereas the opposite has often been the case in the old time cycle(s). We are shifting world realities so we better start acting like we are actually sharing one. How else can the world reflect an improved state of mind? We were never meant to do or make it "alone" here – and neither are we going to - because the world is never greater than the sum total of its individual parts. All human life - the people that inhabit our world - are interconnected. If we are unhappy about something we can start doing something about it where we are at. How often, have we bought into the idea that we are rendered powerless and in this way contribute to a state of uselessness. When we realise that the change we seek must start with us, it gives us power to make a difference.

Becoming Authority

When something haunts us, it is often because we are afraid of really facing or feeling what is. For instance, if we have given our power to so-called authority figures it is only because we have allowed others to tell us what is right or wrong, to do or feel guilty about something. If we yearn for something or someone else than a "Big Brother or Daddy" to survey, watch over or show us the way, it is because we are beginning to realise that we - the people – hold many more of the answers and keys to the resolutions we are looking for. If we question old establishments' ideas or rules during whatever movements, it is because we are intuitively beginning to understand – that now is the time to reclaim our ability to doing things differently that better reflect where who we are, where we are heading or what we deserve.

Real change must come from within or from the power of the people rather than from above or from small powerful parts. If we are looking in the direction of more reliable, responsible, sustainable, self-educative ways of being or acting, now is a good time to do it and go....natural, holistic, eco, green, alternative and possible to replace old or outdated ideas instead of buying into fears or stay safely tugged away, secure and the like...believing that we cannot change anything at all.

There is no way around it if we are to dismantle certain ways that show greater respect for our self, others and all life, including the natural environment of the Earth of course. If religious, educational or other authorities are becoming increasingly ignored in favour of the internet, self reliant or sustainable alternatives, this is one of the reasons. The same goes with economic, political and societal authorities and conducts which have been in a crisis of mistrust for so many years that it will take a complete revival of our (democratic) standards and (political) ethics to salvage the value of our political representations in many places around the world.

We have to re-remember what authority really means. It is about coming from a place of truly knowing what is right and wrong and not what somebody thinks is right or wrong that only serves certain ideologies or powers. If there is no integrity in that, perhaps it is better to follow your own path than whatever so-called authority. Certainly, wherever the crowd is going or being misled does not always reflect the direction we should be heading. We need to be clear about our own points or views before we jump on some bandwagon of popular opinion(s) or conclusions, but in order to do so, perhaps we need to know our true values.

Finding Our True Values for the Future

In the past Sun cycle we may have been obsessed with "externals" or face values to be happy, feel that we are doing the "right" thing, accepted etc., but often this is more about following other people's values or external value systems than our own. Have you noticed?

Rarely do these value systems teach or tell us much about our core personal values –that are the most important to us in relation to how we want life to be. If we internalise external values, we can very easily lose a sense of self or direction. It can push us off course because outer value systems reflect what others consider most important instead of what we truly adhere to. As we are changing both the collective and individual value systems during these times we are being challenged to reflect more truthfully on them. Only we can find and know them through our own personal guidance system. Life is a precious gift we have and we have to know our values to figure out how we will use it because what is valuable to one may not be valuable for another. Values help us stay on course instead of getting sidetracked. It is about knowing what is most important in life for you.

From Outer Knowledge to Inner Knowing

Within us is (all) the insight we often need to reach wiser decisions about any challenges that we face. No one on the planet knows more than you do about your needs, your values and potential, what gives you joy, self respect and love in life. If somebody tells you what is best for you – without confirming with your inner knowing (excluding your ego) it is better to walk away. It all boils down to the following:

.".the Kingdom of God is inside of you, and it is outside of you. When you come to know yourselves, then you will become known, and you will realize that it is you who are the sons of the living Father. But if you will not know yourselves, you dwell in poverty, and it is you who are that poverty." - Jesus Christ

Of course that does not mean we cannot learn things from others, but the whole difference is how we apply this knowledge. Do we come from a place of ignorance or innocence? Do we believe that our inner knowing is not as important as the pursuit of outer knowledge to really know this or that or lead a successful life? Knowledge may be power, but.......inner knowing is very powerful.

If we only rely on (following) outer knowledge, we could easily become misdirected or sidetracked in ways that do not reflect our true knowing about our life and its purpose. Rarely will outer knowledge teach or tell us about that. How often do we use or misuse knowledge we have required instead of using the inner knowing we already have in relation to what is really important?

If we make choices or decisions based on inner knowing we become capable in a way that is appropriate to bridge the gap between what we think we might be able to do and what we really can do. If each and every one of us took responsibility for our inner knowing in this way, we would better be able to respond in ways that really support what goes on in life - instead of a concept or an idea about this or that by acting rigidly in a way where knowledge overrules an inner knowing.

Of course that does not mean that we have to erase whatever knowledge we have mastered or learned, but if we can combine knowledge with knowing it helps us in life quite differently because knowledge can then be enacted in a more inner directed way. If we are used to overrule our inner knowing with knowledge we can reverse the process. It is similar to a relearning experience. For instance, if we have taken our driving licence in a country, where they drive in the left side of the road, and we move to another country where they drive in the right side, we do not have to throw our know-how about driving a car through traffic out the window. Rather, we can use what we already know about cars and driving in traffic in a different knowledgeable way.

Releasing What We No Longer Want or Is No Longer Needed

There are times in life, where we may feel that what we have learned or relied on in the past is no longer something we want or something we find useful to take with us into the future. To get to this point and figure it out, we need to get very clear on why that is and what it is we need to release. It is about being ultimately responsible for our own life. We also need to get clear on what we are personally responsible for. First of all we are responsible for the fulfilment of our personal needs. We cannot rely on others to speak up for us about what we need, because then we will not receive the help we want. Here is a pleasing example: If we are in a job that overworks or a familial relation that stresses us to death and we do not convey that it is becoming unacceptable, then we will just get burned out. Nothing will change, but if we clearly express our situation or needs, this can open doors for the situation to change or be altered. Whether that means a new job or family situation that will reflect what we need instead of having to quit our job or leave our family all together.

The second part of taking responsibility by releasing what is no longer needed is about clearing the desk to be able to leap into more or other meaningful action. Many of us have been content with simply letting others, decide what is meaningful for us. Or we simply did not realize that there was something important outside of our old lives or jobs that we wanted to do or that needed to be done. If we find ourselves in times of change and transition, we might be more in a state of urgency with regard to these things and a stronger determination is needed - if we are to step into these new actions or roles. We have to do that which aligns us with our true purpose more. The time of bringing our true being or purpose into action is increasingly here.

Transforming Relationships

We do not necessarily have to let go of old relationships or partners just because we change. Existing personal or business relationships can deepen or be renewed by reviewing or rewiring them through the values that are held in common. Far too often, however, we can become waylaid by events on the periphery of our relationships– or caught in the cobwebs of commonality. What may first appear as a pressing need for sudden adjustment, can actually be a blessing in disguise. As we learn to read between the lines and look at the bigger picture of a personal or professional relationship situation it can become the subject of further development through change with benefits that far outweigh the setbacks.

Sometimes that is not the case and we can see no other way out, but to leave the relationship all together. Even if we are afraid of being or becoming alone, it rarely happens because new relationships can always develop or form through shared values and trust.

Whether we transform or dissolve our relationships we do not have to lose ourselves through our relations, but can see that every human interaction or relationship challenge, that we face, holds within it the perfect solution or image that can take us beyond past problems or limitations. Every problem holds in itself a potential solution if we know how we got sidetracked through the emotionality of facing the situation or problem and how to get through it.

"The only way out, is through."
- Anonymous

CHAPTER 7

FINDING THE ROAD AHEAD

Bridging the Individual and Collective Consciousness Zones

Whether our time of transition becomes a time of upheaval or gradual change, we need to become more aware of our individual role in relation to the collective. How can we embrace and adapt to unexpected circumstances instead of resisting impending change? One way is to learn to see things from two sides of an equation or shift our focus away from ourselves and see how we can apply ourselves in new and more meaningful ways that are truly needed as we move from an old to a new reality.

Change during a transition is likely to bring us or rather our focus to what is most needed or away from that which is no longer needed. It is part of change that cannot be limited or controlled. Like a hurricane, it comes when it wants to, affects whatever is in its path, or devastates whatever needs blows of the New, the Unexpected, or the Wondrous, as it continues on to its next destination. This is why we need to find ways to adjust by bridging the individual and the collective (consciousness) zones. Winds of change can be both stressful and frightening so we need to find a place of stillness in the eye of the storm sometimes with others who can also render help or give support. We may have to reflect or rethink on what is going on in a zone of stillness. If we are internally still - even while being outwardly active - we will be better able to hear the numerous hints, alerts or messages that are coming to us from within and without. When we anchor our beings in an inner point of stillness, we can align ourselves with a place of Trust and Trueness that allows us to assist ourselves and the collective as well.

The more pressure which is on us for whatever reason, the harder or more difficult of course it often becomes to see things clearly from the point of view or perspective that is really needed. If we are to stay clear-headed and make the right judgments and decisions between a multitude of (unexpected) events and possible choices, we need to wait before jumping to the wrong conclusions. These decisions and actions must be based more on truth if we are to make the jump from the old cycle.

Reflection on the Old and Integration of the New

To understand this we need to learn to reflect on what is occurring individually to figure out what it means and what is really important collectively as well. (Self) reflection is a powerful mind tool and key to this, that can help us learn from our experiences before we jump into knew actions and experiences on a lifelong journey. If we use temperance it allows us to think twice while diving deeper into our thought processes about what happened so that we can use a situational appropriateness that enables us to choose wisely among different future scenarios or possibilities of action. This aligns our inner core more with the right things at the right time. In this way we can better make sense of in between times as suggested by the cyclical nature of the Mayan Calendar. Reflection on experiences can happen instantly or over a period of time in relation to our thought processes and emotional states of being. It is about learning from what has happened in the past to figure out how to act in the future.

Because everything we do in the present starts with a thought or an emotion, by reflecting once or twice on this intersection, we can circumvent false beliefs, ideas or assumptions that prevent us from reaching insight we need to make in order to reach the right conclusions about decisions or actions related to the situations and times we live in. In this way new ideas and thoughts can germinate within our consciousness to open up new possibilities.

We have all tried what it means to reflect on ideas, experiences and concepts through our personal and professional studies and the like, when we think more deeply or closely about something. It allows us to learn from experience and save ourselves and others from future problems or useless efforts.

In pivotal, transitional times, where a multitude of things are happening or could happen out of the blue, we need to stop and think for a moment to reflect on what is really going on. With so many different scenarios and options, reflection becomes really important to integrate and cope with change - particularly during 2012 and beyond.

It teaches us how to become more mindful about our human actions and interactions. It can help us remember that whatever happens can be a valuable or useful lesson no matter how good or bad it seems on the surface, so that we can move forward from that experience.

By integrating the past with future possibilities through reflection in the path or reality ahead can become less fragile as we learn to merge the two. It gives us the privilege of using our flexibility and adjustability with the incoming "flow" of potential future events before they might happen.

> *"Your task is not to seek for love,*
> *but merely to seek and find*
> *all the barriers within yourself*
> *that you have built against it."*
> *~ RUMI ~*

Shifting Inner and Outer Coordinates by Following the Compass of Our Heart(s)

How can we find the new coordinates that allow us to navigate into the future with more confidence? How can we act on an emerging new reality that enables us to find the right solutions?

As we progress through 2012 and beyond we need to remember that the future is not fixed: Like the needle of a compass the road has shifting or fluctuating coordinates. Even if we find it hard to figure out where the North Pole is and which direction to take there is no need to give into fear. By trusting our hearts we can bridge the immediate gap between the past and the future. Believe it or not, the heart is the key because it allows us to filter out or recalibrate things in our life. We are all connected through the core of the human heart. If we can feel and understand it the separateness that we sometimes feel and may experienced then becomes an illusion because each of us is a part of humanity. All we need to do is remember to align with the heart that is perfectly present already. When we do, there exists no separation between us – only the one we perceive. So the question is what are you/we paying attention to? The separate being or that we all belong to the same human tribe? To get to this level or state of consciousness we need to stop looking at all "the stuff" in our life which does not appreciate us or we do not like and re-orient our focus or perspective towards the everything that we belong to which feels good or tells us more about how wonderful we are – rather the opposite.

How can you become more aware of the essential beauty that you are? Look for ways and you will find it because you already know it. Look for ways to be impressed with and love yourself and others as they are. Although this may sound airy fairy or like a naïveté, centring ourselves in love in this way works well in life if we embrace and understand its core meaning. We have to learn to co-habit the planet by respecting our different juxtapositions and differences in order to resolve our personal and interpersonal conflicts to evolve and move forward together. How else can we/you find the path ahead if you/we are not heart centred? The true path goes through the heart because the heart bridges the gap between past, present and future realities. It links the lower (emotional) aspects of self with the higher, more illuminated ones. It bridges the individual and collective differences if we chose to embody its luminosity. Even if we find it hard or difficult to love something or someone we hate or dislike, aligning with the heart can help us feel better about the situation or persons involved because love neutralizes whatever we are unhappy about or whomever we are unhappy with. We all need love and desire loving relationships – although our worldwide definitions of them might look different. If we can see that the heart is the key - the only thing we absolutely, really, always and truly do share - it becomes more difficult to lose our patience or temper with people(s) or situations we are angry with or unhappy about. Initially this may seem like a difficult concept because we live in the "real world" – but if we cannot find a way to love in the midst of personal difficulties or interpersonal hatred, how can we ever get or move beyond the entrapments of the lower emotional state to a stage where we become more spiritually liberated?

It does not really matter if we are Muslim, Hindu, Christian or Buddhist – we are all humans and our heart is the organ where love is a feeling we all share in spite of our religious, social, economic and other differences. Love is the key to what lies behind it all and all there is – it is the only thing that is real. It is the glue that connects or binds us and the path to creating win-win situations that reflect parties' need for loving resolutions. We cannot move away from conflicts through separation, strife or war. We cannot learn what we need to learn from our differences or different points of view without embracing ideas of love. The path of life on an evolving curve is invariably tied into or with it. Somehow along the way it becomes the only way to move upward or forward. Even if we feel angry, sad, unsettled or separated, we can only go so far out on the line of on an unloving, dual path before it eventually breaks down or dissolves.

Love is the only reason why we are alive. This goes for the love of our planet, human relations, family, friends, animals and what else? What about our organizations? There is no way we cannot tie love into it if we are to function in a group or an evolved society. So we need to "think again" if our ideas about love are not congenial with the ways we organize ourselves and our human lives in society. It is the essence behind how we can resolve whatever problems, difficulties or differences we have – because once love arrives, weather we are rich or poor, workaholic or unemployed, in charge or feel powerless, we can see that love is such a force and part of our human being – that it allows us to "purify" our relationships by "un-separating" them. Love is not "dual" by telling us to do or die – like or dislike. Rather, it is a basic human value, principle or ideal that we all share and can learn to embody if we want to function and embody the oneness of life.

Often, we block ourselves from this love, literally and/or figuratively and/or emotionally. Sometimes consciously, owning your fear of commitment, fear of loss or fear of betrayal; sometimes unconsciously, sabotaging things as soon as there is any kind of conflict, real investment or compromise required.

One way or another, once we come to the conclusion that we need love to keep our sanity and respect for each other intact instead of apart, the foundation of our being is altered and no matter our position or juxtaposition in life, we can start to apply more of it to reach out to others in personal and professional life.

If love becomes real in down to earth human terms, it can perform miracles if we allow it to. Just see what happens when you tell someone you really like or love them. Do they avoid you, become surprised or not light up and smile? In spite of whatever disagreements we might have with one another – love comes from this other or universal place that tells us the truth about our unity in duality which we might need to live through, resolve or transform things for the better.

Just see what it is like to listen to the words and phrases of those who embody the kind of confidence and light-hearted energy which you desire to experience. Is it not better than the opposite? If we desire the experience of living as real human beings, we need to be mindful about how we can love self and others in a confident and asserting manner. Love activates the reality of our soul signatures when we learn to live and share our inspired visions with others.

Until we learn to understand LOVE in this way, we cannot expect to experience the gifts of love that truly drive and support us while we are beings here on Earth.

Aligning With Our True Being in Action

It is important to continually remember that although many people are feeling the Shift that is taking place, many are not consciously aware of what or why it is happening. How we feel, experience or live through the Shift vary greatly depending on who and where we are in life. It depends on our past development and the extent to which we are consciously able to work with the stream of energy that is inundating our Planetary Sphere. Once we surrender our resistance during the Shift, however, we can step onto a bridge that will take us to our next stage of evolution and into a new reality. It is about giving up ideas about how life is "supposed" to be or what it looked like in the past. We do not have to repeat the mistakes of the past. Also, the ideas of individual struggle makes matters worse when what we should do is connect with each other to experience the Wholeness and Unity of Life.

By replacing our egos with a sense of humility, ignorance can be replaced with curiosity, innocence and inner knowing which opens our hearts and minds to the humanity of which we are all a part. As the 2012 Window of Changes is blowing out the barricades we have created in our minds and tearing down the inner walls created by emotional fears, it is trying to show us the universe's Grace. Surrendering to its process can become an impetus for lasting change that when we look back we would never have lived without. It gives us the opportunity to discover who we are, what our potential is or what life is all about. The process is as much about ourselves as it is about connecting with others as we leave old ideas behind and reposition our selves towards new ways of life.

We can learn to bring our one, true being into action where and when it is needed by offering our love and support as we become loved and supported as well. Sharing and offering our service through acts of love opens up many opportunities and even solutions to the challenges we are facing. Working together in unity can resolve many of the personal and global issues we have been struggling with in the past. It can teach us how we to create a new more liveable and sustainable planet that can benefit us all for many years to come. But before getting to this point, perhaps we need to take a closer look at how we really deal with change as we open ourselves to our true potential associated with the Shift itself.

The Great 2012 Window Changes of Opportunity

Regardless of your age, status or situation in life, there has never been a time of opportunity like this one to search for personal meaning and to know your place in the world. It is increasingly occurring amidst the greatest cycle of changes we have ever witnessed. Is it any wonder if we are curious to know more about it in relation to who we are, how we fit in or what we can do to embrace change in relation to the big picture of a steady stream of changing global or personal events?

Perhaps you need to ask yourself what you can do to find a sense of fulfilment or purpose in life. How you can reconcile yourself to issues and make friends with change? The following is only a background that can help you understand how to deal with what is happening in order to move forward with greater understanding and self – empowerment.

The Dance of Creation Involves Change

The dance of creation naturally involves change. Our parents created us and we are living and breathing that dance, even when we think that we are standing still in life till the day we die. One of the differences now compared to previously is that you can become more awake with more conscious awareness than you may have had before. The ability to see things more clearly as they happen may become more acute as we become better skilled in accessing our intuitive or inner knowing and also sense things more profoundly. Keep in mind when or if this happens that it holds many new options for you. There is not a particular or exact way – to open your self to the faculties of the right side of the brain. Do not allow yourself to go into fantasy or fear about it either. Let go of expectations, too. When you expect something or someone to act in a certain way, you limit yourself and what you could experience. You set yourself up for disappointment or potential suffering when something else could arrive that might be better.

Oftentimes, if we are out of balance or involved with self destructive emotional or fearful behaviour, we may sense its potentially unpleasant consequences. As you are responding to whatever information, factor in the possibility that some or all of what you are hearing, seeing or sensing, could be your fear, anger or ego talking. It takes a developing skill to discern whether it is your intuition or your fear communicating, your ego or your true self. Discerning fear can be tricky, because the ego has many disguises, as you might learn over time. If it is really your intuition "speaking", there will be no scare or manipulative tactics involved or needed. You will simply be given information or shown some ways you can change yourself or change course in alignment with who you are, where you are. You will be guided to different choices, situations and people that better reflect where you are coming from or where you are supposed to go. When you act on those choices, you set in motion a different and more positive future where you are in the driver's seat of change rather than a victim - in ways that can benefit yourself and others.

Heightened Sensitivity

What also occurs as you become more awake is that your sensitivity heightens. In increments, as you progress on your spiritual journey, you may become aware of more and more. This awareness is sometimes a knowing, at other times more like a feeling, and sometimes both. What does this mean? In general, you become more sensitive to energy, including that which is within and outside of you. That means you can more easily get in touch with your own self – your joy as well as your sadness. It means you are more able to sense when others around you are under stress, feeling angry, or trying to manipulate you.

Being more awake comes with the responsibility of being willing to face what you see. Some of it will be pretty. Some of it might be ugly. Whatever you encounter might stir a plethora of feelings, ranging from delight and joy to fear or repulsion. A vital part of an awakening journey is becoming conscious of your true feelings. In order to progress, you will need to learn how to intelligently work with emotions that are a natural part of being human. Masking your feelings through unnatural means will only impede your natural spiritual, evolutionary growth.

You cannot be numb to yourself and your life experience and become more awake or enlightened. You must be willing to know what you feel, transform negative feelings, and avoid acting them out in the world. This is (y)our responsibility. When you act on your hate or other negative emotions, you harm both yourself and others. It is a real energy force that you send out into the collective experience, and like a boomerang, it might come back to you at some future time.

The Quickening

Another difference is the acceleration of energies that is a part of this Great Shift of cycles. The quickened pace of evolutionary changes can be mind boggling because there simply is no previous modern day reference point for this magnitude of shifting. Even wisdom keepers are evolving, and as they do, they can offer updated insights that can help you to take the next leap in consciousness.

While so much is shifting and changing (within and without), although many things in the physical world are the same, your great-grandparents would most likely have a hard time keeping up and recognize how little of our modern lifestyle, resembles their old way of life. This fight with change between generations remains an even greater key stumbling block for humanity today. It is almost like our biology is equipped for the Stone Age while our mind is blown up with IT-technologies. The lack of a spiritual understanding or mindset does not make our adaptability any better while we have evolved technologically far too quickly for our spirits to keep up with that.

Stretching and Moving Beyond Our Comfort Zone(s)

Consider for a moment, a time in your past when you broke a leg or lost a job. How easy was it for you to come to terms with the change in your life's routine? What did you do to adapt? How did you find your centre while you were healing? Did you find other work? Adapting to changing external circumstances often means moving out of our comfort zone. It is often not nice. We like what we are comfortable with, so we gravitate towards people and situations we already know. We do this with our feelings too. The habitually angry person sets up a cycle of scenarios that fuel more anger. The habitually depressed person looks at life through the lens of their sadness, so they continue to feel depressed a lot of the time etc.

Surely, if you knew that you were creating more anger by focusing on your anger, and you got tired of that, you would be inspired to do things differently. You would seek solutions that came from your essence or inner source that would lead to more peace and joy. You would move out of your comfort zone long enough to find sanity and a better way to be without sadness or anger. New positions and directions are not handed down, we have to choose them. If we allow it, our core or inner being often knows how. It is more adjustable and adaptable than we often think with our ego mind or from the realms of our personality. During the Shift, what is increasingly being revealed is that if we keep repeating or creating the same life stories, life might be pushing us beyond these personal comfort zones. To embrace these spiritual changes requires that we take risks in order to follow what we know to be true - instead of playing the same old records.

Change Is Constant (Movement)

Change is constant movement. It can appear slowly or quickly. It can seem like it is not happening at all. You may feel that things are changing too fast or you may feel stuck in limbo. You may be frustrated at your seeming lack of progress, noticing only what is wrong. Or the opposite may be the case. You may feel like the movements of change are happening outside of your control, especially when you are not tuned into the subtle ways that changing energy moves.

To go forward more skilfully, you may want to develop some new ways of looking at yourself and your world. To begin with knowing that you are experiencing the changes means, each time you experience something, you are an integral part of the change. You need to pay attention to it, like an amateur chef without a recipe, throwing things into the pot. Using your common sense and taste preferences, you do the best you can. Your habits also play a role, reflecting what types of food you typically like. Your family's conditioning impacts you too. You learned to like foods that you associate with comfort and love. As you are creating your meal, you can either do so consciously or unconsciously. In each moment, your creating will take one of these forms. Sometimes you may be very aware of what you are choosing to put in the pot, selecting each ingredient with care and attention. Other times, you may have your focus on other things, retrieving the same ingredients you used yesterday or the day before.

Sometimes you may lack a key ingredient or have too much of one thing and not enough of another. You may even decide to toss out what you made and start over like the Swedish chef in The Muppet Show. When you create or change something, you may either be pleased or unhappy with the results. Do you ever feel like this in life? Do you wonder what key ingredient you are missing? Do you question how you can more skilfully create or change things in your life?

A Checklist for Embracing Change during the Shift

The following is a checklist for embracing change from the driver's seat during times of change. Particularly the ones many are going through during the Shift. You can work with some of these ideas as often as you feel guided, inviting your higher wisdom to nudge you when you get off course. Or tell your friends about them or share experience.

1. Live life each day as fully as you can... How many of us do that? How often do we live as if this day would be our last? It is about truly being grateful for the opportunity of being alive. Remember that you are an integral part of Creation's most auspicious moments....You are not simply a witness, but take active part in the Great Wheel of life. Living through this window of time-to witness and experience some of the greatest change is quite something. Know that everyday is important and decide you will take an active and conscious role in life.

2. Monitor your thought processes... Keep in mind how powerful your thoughts are in creating also what happens next. Thoughts are things. They may be invisible, but they always have an impact on your mindset and consciousness. This is true whoever you are. You are in a continual process of creating with your thoughts and words. Decide that you will monitor your thoughts with love and higher intentions. This means having compassion for yourself and others even if or when the road gets rocky. It means letting go of the need to be right, to criticize or to judge. No one is really perfect, we are all here to learn (to use our thoughts wisely) - yet there is a Divine perfection in it all.

3. Remember that there really are no accidents... Look for the higher purpose in things. When you cannot perceive this directly, trust that things often are not as they appear or seem on the surface. Learn to look more deeply beneath the surface for answers than you did yesterday. It is about questioning more from your heart to be able to act from a grounded sense of intuitive inner knowing. Remind yourself that you and others are evolving, and in doing so, you can learn from everything around you. Avoid the temptation to become prideful and think you have all the answers.

4. Decide that you will see what is there.... Pay attention to what really goes on. You need to learn to look deeper than surface appearances, beyond what is presented, what is announced as fact. These are times of increasing transparency, but it is our consciousness that will determine how or whether we see things to receive the right information. Not everyone will perceive events the same way. Those who are not ready or do not want to see the truth will have their own version. Do not feel you must convince them or impose things on them. Stay in your centre, and trust your own truth. In the eternal realms, no proof or manipulation is required. Truth simply is.

5. Place your focus on what really matters... Stay on track with what really matters throughout your day by learning to live as consciously as you can. Develop discernment and self-trust in your inner wisdom that will show you, step by step, what to pay attention to and what to disregard. Attention can teach you many things about the meaning behind life's events. Noticing how you feel and think in relation to situations and people and how you are responding to what happens in your world is important. If you are feeling unhappy for some reason, allow yourself to acknowledge that and figure out how you can anchor the feeling of joy within your heart. Getting in touch with what brings you true joy and gratitude will give you more of that. When you find that you are feeling angry or fearful, inquire within to discover the source of your anger or fear and work intelligently with the feeling, so that you do not allow a person or situation to poison the next thing you think, feel, say or do again.

6. Update your relationship with linear time... Separate out your need for order and your need to be fully present with life. Your only true power to create is from the now. Let go of your war with time, with deadlines, and prioritize your growing to do list. Acknowledge that everyone gets baffled by time sometime. As you become more aware of your relationship with time, time takes on a new meaning. Of course our clocks and calendars are convenient tools to keep track of time. They provide a practical way or measuring meetings and arrange the details of your day. Beyond these practicalities, the use of linear time is relative and not helpful for living in the present moment or figuring out your psychological time. In one instant, as it occurs, your life could change. It is only from within time you can make certain personal, psychological changes. Calendars are just measuring sticks, which help you to track the seasons and cycles.

7. Become friends with change... There will be lots more of it. A great quickening is now in full swing, with more of humanity awakening at a rapid pace. You are an integral part of that process......shifting is happening for and inside of you too. There might be so much you want to change about the world you see, but the most productive way to initiate those changes is to focus on being as loving and kind as you can be. If you feel there is not much to change - include yourself in the change equation! Whatever challenges you face should become the energetic fuel on your path of change. When faced with obstacles you can transform them, by becoming a luminous gem in the process that you can share with the world. Rather than complaining about change, express gratitude for the possibilities change is instigating.

8. Keep 2012 and other time points in perspective...The shift is occurring in a large open window around 2012 and is more of an ongoing process than a single time event. Evolution is unfolding with increasing momentum during your lifetime. Whenever you see something ends, something new can begin. It is like great historic events which represent an enormous potential for humanity to participate and to co-create needed changes. We came not for one date in time, but have a front-row window seat as the planet moves into a higher vibration. The movement into higher levels of consciousness will not always feel comfortable. Sometimes we will feel dizzy or simply tired or want to cover our eyes to avoid seeing what is there. Sometimes we will question whether we can go on or not. Do not worry. The eternal you has no doubt that you will succeed.

9. Avoid shutting down your senses... Both your physical and intuitive senses are integral to your movement into a higher vibration. You do not want to numb them with chemical substances, overwork, a non stop diet of sugar or mindless media input. Your body is the Earthly temple for your soul. It will give you vital information about what is important, what and who to trust, and what you need – even if out of balance. Opening and trusting your senses allows you to connect with your heart centre and all life more often, allowing for refuelling and reflection. Allow the insights you receive from that place to guide your next steps. Avoid obsessing on the minutia or self-pettiness which often dominates ordinary, modern human life.

10. Watch your emotions and believe in yourself... Emotions can be your greatest helper or your worst enemy. Learn to identify how you truly feel, when you feel it instead of letting your emotions run away with or control you. Become intelligent about how you respond to what you feel. Decide that you will use your emotions as a beneficial tool for your growth, with harm to none including yourself. If you realize you are fearful, anxious or depressed, address these emotions without judgment. Remember that anyone can feel negative or bad things. The key is to find solutions to them - so you do not go into self-sabotage, pitiful or panicking mode. To every emotional problem there is a solution. Underneath whatever fear or anxiousness, when you go deep enough, there is some kind of power or joy! Your genuine self contains joyful, happy qualities. They were naturally born within you, so that you could learn to open yourself to them. The more you intentionally work with your (negative) emotions, the more it helps you to flower into your highest potential.

11. Simplify your life... How many of us fill our lives with clutter till we cannot move or breathe or become all cramped up about stuff that we have to worry about? If finances run short during the Shift how will you be able to take care of these things with a pile of bills that you do not know what to do about? The more spiritual you become, the more you realize the most important, yet beautiful and fulfilling things in life are not the things you can buy. When you pay attention to the inner you, you will find that life is really quite simple and that the world does not have to rule your spiritual life if spirit or nature makes your day. Simplifying your life gives you more flexibility by allowing you to change directions and move swiftly whenever you chose or need to. That is why keeping your life simple can bring you back in the driver's seat.

'The world is won by those who let it go!'

- Lao Tzu, Father of Taoism

CHAPTER 8

JOURNEY INTO WHOLENESS & JOY

The Shift Might Squeeze Our Humanness in New Directions

In the midst of this Great Shift, we can see paradigm shifts, stirrings or changes in our current ecological, economic and social predicaments occurring all over, already. Depending on how sensitive or open minded we are, we might feel a sense of pressure is in the air, or see a need for individual and global changes. It means you are not alone. At least now you know, the Shift is occurring for all of us all over the planet and we need not look far to be receiving strong messages about some of the things associated with it that need attending to. Wherever we are, we need to stay alert and encourage ourselves to pay attention and listen to what is really going on.

On a human level, we may at times feel lost and disorientated, because for a while as the remnants of the old cycle are cleared away this can leave a large and very empty space, with nothing immediately obvious to take the place of what was there before. Almost every area of life can be affected. For example, we may want to review social or family relations that have drifted apart, existing partner- or friendships where the finest days may be over, career ambitions and interests you have pursued so far that no longer hold the fascination they once did, businesses or associations closing down. Inevitably, a metamorphosis does not pass without making its mark. Our outlook and philosophical perspective go through a corresponding phase of renewal, as we come to terms with the essence of regeneration and where it might leave us afterwards. If a major cycle is concluding, at the same time a new journey can begin.... This will not happen overnight and there could be several false calls, dawns or starts.

Therefore, rather than preparing for some future moment(s) down the road, let us notice how we are participating in life at *THIS MOMENT*. Rather than focusing on an imaginary destination, we need to realize where we are in our personal or planetary transformation process of finding more joy and wholeness.

The 2012 synchronization beam is really a call to purpose. What is *YOUR* life's mission? What gives you joy? What is the unique role you can play or path you can take during these times?

We need to follow our inspiration and urge to find our passion, so we feel alive and directed. Following fearful thoughts or paths will not serve us, because the voices of self-doubt and anxiety make us a part of the problem rather than inspire us to be part of a solution. They clog our vision or lead us to shut down our senses so that we cannot see the positive potential or use our creativity in relation to reality. What we need is to feel engaged in life by knowing we are part of a collective being that is moving towards a new future. How else can we feel excited about life?

The wisdom of the 2012 Mayan prophecy tells us to look directly at our lives, here and now: Where is our focus? What are our guiding principles and motivations? What did we come here for? What would we like to do or create? How can we live in greater harmony with all our human relations? How can we feel more truly alive? How can we embody genuine compassion? How can we face our edges and grow beyond what we thought possible? How can we become more aware or reflective? Can we deepen our humility and share our talents and gifts to influence our culture and help manifest a better world? How can we be of service and learn to contribute to the Whole? How can we work on behalf of future generations and cultivate balance through our ways of living with receptivity and action?

On this diverse planet, we all have different strengths, abilities, callings and sometimes their opposites. Rather than look to the world to show us our rightful path or what to do, we need to look deep within also to navigate these times of the great unknown.

It is essential we learn to hear the voice of our inner wisdom. No one can give us permission. It must dawn and dwell inside of us and be born of our own direct experience. We each have our direct connection to our Spirit so we must keep learning how to hear our divine call, intuitive directives or whispering within. Being spiritual is associated with the ability to hear the voice of our inner guidance and perhaps it is the most practical skill we can cultivate in these times as it will inform us to how to synchronize ourselves with the right place at the right time.

This process of synchronization with our intuition and the universe at large is greatly amplified by learning how to live in harmony with the natural time cycles. This awakens our conscious awareness of synchronicities to the natural order of life and others. It can practically be done by choosing to follow the harmonic mathematics of the Ancient Maya, purposely designed as a solar-lunar-galactic calendar for the modern world to reconnect with the sacred essence of time. In light of the 2012 prophecy, many people around the world are already choosing to synchronize every day with the new wave. As we follow the natural rhythms and cycles of this new paradigm or era, we can be guided to infuse every day with a unique, spiritual focus. If we do this together we are forming a new web of unification, and becoming conscious vessels of the One.

The One is our Creator or Creation. We may not see or think of it this way, as some would rather call it God or religion, but we are nevertheless all part of it in some way.

There are always "fears" or a ga-jillion reasons and excuses not to trust or follow God, but if we listen very closely to that voice within, we can see that it brings many opportunities. Life in the new paradigm is for those who are willing to "step up to" and be in that game...

As the shift of realities or paradigms continues, we may be loosened from old beliefs or settings which determined our ways of thinking or living in the past or the old cycle. This frees us to move into new, potentially more rightful positions that can only be found by shifting directions if we become courageous, open, willing and (cap)able...Does it seem or sound like too much?

As the Great Spirit will have it, new ways of being or ways of life must reflect a new growing higher awareness that will soon become more available to us as we pass through the Great Shift. If it means taking away ways of our ego it is only because the time is over for hiding or tucking away our true being.

If we can courageously open ourselves to embrace these new spiritual directions or pathways they can open up in ways we never thought or dreamed possible, yet might have known somehow or somewhere was actually meant to happen for us. Why is that?

In the old cycle we explored many aspects of our physical, emotional and mental selves through the ego's lower modes of action and reaction, but as we move into another cycle – a higher ether element is added to the mix of the physical, emotional, mental and spiritual aspects of our human being. It is a fifth element that weaves the other elements together at a higher level closer to Creation or Source.

Willingly or unwillingly, we are in some way being "pushed" to release lower aspects of self which contain less of the spiritual life force or energy that permeates all things, so that we can bridge the physical reality with the universal or spiritual. An analogy is when we go through the experiences in life. When we are born, we needed physical sustenance like food from our parents. As we grow older we needed emotional support as well to develop our social skills and emotional abilities. When we became teenagers our mind expands and evolves in order for us to learn and develop our mental faculties. When we become grown ups and middle aged we start to apply all these faculties and as we grow old we get to see the experiences we have had in spiritual retrospect. And finally before we leave this world we hopefully go through and see all the experiences we have had in life in a whole new light. Isn't it interesting how a human life can contain all these elements and that everything has a beginning, middle and end before it returns to become part of the great spiritual cycle of life?

Life gives us many opportunities and experiences through the various levels of our human existence, and understanding them all from a spiritual point of view takes time, effort and patience when eventually we have to turn to our Creator, the Great Spirit or God. This is the spiritual aspect of life. It can also become our individuation or self realisation where we come to remember or rediscover who we really are in relation to all things. The Great Shift gives us an unprecedented opportunity to come to some form of realisation about these aspects of life in relation to the cosmic or universal.

If our interaction with life in the old cycle has taken us away from our true spiritual self or connection to God, through the experiences of the previous lower Mayan Underworlds, the Shift could lift the veil that prevents us from seeing it more clearly. It could show us more about who we are by making us aware of how our lower human self through familial conditionings, societal and worldly influences and the like that have covered up the vision of our true inner being. When we look at society for instance, we can see how certain institutions, structures or systems have been created from a limiting or debilitating understanding of our human potential – which does not support our life or reflect an evolving human civilisation. Our societies and histories still reflect the previous Mayan Underworld stages as we were struggling to survive through our past evolutionary history. It was a gradual, slow moving development or wake up call which has taken us from where we have been to where we are now and going next on our human evolutionary journey.

As we pass through the final days and nights of the universal Mayan Underworld and eject passed 2012, we are on an accelerated wave of change, and path of awakening and new beginnings. We are only slowly beginning to understand how we can integrate old structures and belief systems with new ones that are more meaningful, less painful and allow our true possibilities and potentials to emerge.

The past Piscean Age had an element of "human sacrifice" whereas the new Aquarian Age is more about sharing the love, truth, wisdom, unity and universal gifts that exist within our greater community of joyful hearts. We are in the (slow) process of emergence where many of the structures and systems we have created in the past are dissolving before our eyes in order to give way to new ones. Let us face it!

We have relied on the past, but as or if it becomes less relevant, because we are changing time cycles, we need to find a new trust and faith in the future as we learn a higher way of being and living in consciousness.

Once this spiritual insight or knowledge becomes more of a layman's term – and seems less intimidating, new spiritual concepts and ideas will become more of acceptable as the norm.

If we are to make it practically useful, we need to ground the "new human" or "spiritual consciousness" with new ways of life. Essentially it is about (re)evaluating which of our old beliefs and parts of our societal, governmental and social systems that are serving our human ways and interactions in the long run. Do they lift or bring us down? Do they take us back to where we were or move us forward? Can we live with them? How long? And so on....

If we find ourselves at a cross road or choice point with regard to notable institutions or establishments, is it because we are "shifting" into a new greater understanding or higher perspective on things to be able to release our future potential. If important changes have to be made in our societies or communities on the basis of new insight and knowledge, so be it if we are to progress and make personal progress through the Great Shift.

Personal Revisions & Reconfigurations of the Past (Cycle)

Many events in our lifetime take us through cycles with a beginning, middle and end. Events or cycles are not always orderly or obvious. Sometimes the end can feel like a new beginning or a beginning feel like something is ending. This is because many events in life and often overlap. If this sounds confusing, it is, but where we are in a cycle of life is often less important than what this cycle does to us – right?

If our youth slipped away it may be because we did not feel like exploring how to be young or we have forgotten what it was like feeling youthful. In whatever way we go through or experience through life's cycles, it bring us to a point sometime where we can integrate what we have experienced through understanding that which becomes our wisdom. This separates linear time from cyclical time and prepares us for the next evolutionary cycle.

Along this time-shifting process, we might revise or reconsider previous decisions and directions made based on the old cycle or paradigm and its value or belief systems, before we can figure out what they meant and how to proceed or where to go with the next cycle. As we find ourselves at an intersection with regard to these things on both a personal and collective level during the Shift, the future cycle and its pathways are not necessarily made clear. Perhaps or rather, they are only gradually being revealed to us - through the twists and turns of its curves.

If this sounds a bit cryptic, just ask yourself if you have considered breaking with or moving away from something or someone in your life that used to hold your interest or give you satisfaction in the past, but no longer feel or mean the same as it or they used to do. Or if some of your friends, members of family or other people around you are feeling increasingly stressed or challenged to make a change to do something different about their life. Do certain systems or structures in our society that we used to believe in or rely on, no longer hold the same significance, reliance or relevance?

It is a natural part of life to change cycles. We do it when we live and die and possibly live again, only this time it is different because we or the world do not "die". We are in a way forced to make revisions or change cycles while we are still here, so the individual and collective "shifting" can be dramatic, profound or drastic because beginning a new cycle allows greater difference or bigger choices than the ones we may have experienced or made in the past. Do you see? In spite of how odd the idea of a shifting cycle or reality may seem it does offer the opportunity to move on or get on with change more quickly - in relation to who we were or where we have been in life.

We are the ultimate creators of human life, so isn't it about time that we become honest about it? Instead of playing victims of a changing world perhaps we should look at, how we can create a (new) world that better reflect our true human nature and potential and with it the possible contribution we came to make at this particular point in our time for future generations to follow?

If we do not feel or see a need to make this happen at all, is it because we are too absorbed with old personal agendas that leave no room for change?

There is always the possibility or likelihood that we regress into the past or a path if our consciousness becomes clouded by hidden emotions or we simply chose to stay rigid and "stuck". Of course there is no final judgment day because life and evolution continues even if choose not to participate. In other words there is always going to be free human will and choice for whatever personal scenarios.

Making no change is also a choice. The universe does not care or judge. It simply awaits our human decisions while we slowly or quickly move forward with our evolution. It is all a matter of spiritual possibility and personal choice.

Revisions of life can only happen when we are ready. Anytime, anywhere it can involve people, situations or places that we have once shared our life path with when we are ready or able to resolve them.

Similar to a revision, when we are in the process of making a change, aspects of our self or life may suddenly be brought to the fore that need working on for some time before we can fully resolve or release them. We are all participating in this great wheel of life and as we move forward we do so at our own pace in time. A shifting of cycles makes no exception to that because we are always moving one small or bigger step at a time.

If we feel we are moving beyond an old aspect of self or experience - zooming ahead with a new decision, direction or path in life is not necessarily the easiest or wisest thing. Change always needs a certain time to germinate as we are in the process of revising, making or winding certain things up before we can move on. It can be more or less pertinent or immanent. It really depends on who we are and how far we have progressed with issues on our life path.

Either way, we can feel settled or unsettled, stressed or de-stressed, tense or relieved, clarified or confused by whatever change processes that are taking place. If we are evolving out of who we were or what we used to do we are moving towards a new place or point that needs to catch up with us from the future before we can move forward. It is almost like the future is moving towards us before we can meet up with it and arrive in our "perfect" spot.

Whether we are aware of it or not, recalibrating or reshuffling life backward-forwards through the present can be quite an experience which can even lead us to a point where our higher self suddenly comes to new realisations stemming, of course, from our experience of life.

We did not come down here with a scroll from Heaven telling us what to do or how to achieve some kind of illumination in relation to what we came here for, so oftentimes we are in a trial and error process where the necessary adjustments with our lower human self need to be made.

This allows us to slowly, but surely figure out more about who we really are, what our potential is, what we truly want or what we came here to do before we plant the seeds that allow us to go out and experience it in the world.

Without a road map or manual, life can sometimes be both a bewildering or exhilarating experience depending of course on how we chose to look at it. No man or woman is certain to experience this or that same thing, and we sometimes make the "right" or "wrong" turn when choosing a direction. That is both the beauty and the ugliness of being in a human body here on Earth because how else are we going to figure this life out?

> *"There is nothing in a caterpillar
> that tells you it is going to be a butterfly."*
>
> *~ BUCKMINSTER FULLER ~*

The Re-birthing Process

When we spend a life time what happens at the end of the road? Does it mean that we can never come back again? Are we forever gone? Is it like the passing of time that once the clock stops there is no way we can reset it and have the seconds, restart the minutes and hours again? No of course not. The end of a cycle is not just simply the end – rather it means that we are coming to the end of a road we have been travelling and we are making a turn to begin a new one. It is a bit like a caterpillar – once it leaves its shell and unfolds its wings it allows the butterfly to fly. What did the cycle in between change? What have we learned or what do we know now that we did not know or know how to do before? Are we prepared for our next birth or flight?

Of course this is a simplified allegory, but it explains the experiential realization on a journey that suddenly reveals a path which is closing behind. It is about standing alone with an old trajectory or desire to go in a certain direction which is now over or that we need to move away from in order to push or move forward - and where does it lead us?

Indeed, the questions asked on our life's journey, may be important particularly the end of a road where we can feel stuck, in doubt or unsure of where we want or need to go next when an old journey or cycle is finally over. We may even think that coming to the end means that we have not made any progress, but that cannot be farther from the truth - because the end of something is often necessary to forge a new path ahead. Any part of our journey is important even if it did not provide the results we thought it would. After all where would we be or have been if things did not have more than a beginning and end? How could there be any life situations for us?

Any area of life that we have agreed to complete at this time sets the stage for a new cycle. It sometimes appears within our relationships, health, work, career, successes and failures, finances, children and parents, siblings, friends and our common place in the world. We cannot experience a cycle without going through it and any one of the areas or points of a cycle – good or bad - teaches us something through direct experience. While else would we have come here?

Even if we have no clue why we had to go through it, we can still see that something happened and it might reflect the inner emotional states that we were carrying at the time of the experience. It all boils down to the ways we process these experiential energies. Since every aspect of our life is connected, we cannot isolate each part from the other areas; be it emotional, mental, physical, spiritual or whatever. Working with a single area - will often be expressed through many different levels and situations in our life time.

Cycles begin in childhood, which is when the foundation of our life is established before it continues throughout life. Everything we do is related to some aspect of a cycle and we unconsciously gravitate towards the experiences that will help us complete it. Through ebb and flow, it pertains to relationship we choose, choices we make, opportunities and elements from our field of potential – manifest or not manifest – all are related to expressing the energies of our life cycles.

The re-birthing process occurs - just as an old tree leaves a tiny seed for a new tree in the future there is always something new we can look forward to. It is our movement into a new, different level of experience and understanding that is our achievement, not the end itself.

How do we know we are "there" then and have learned what we needed to learn before we can be "re-born"? Often we do not know whether we have succeeded or not until we actually move into another cycle and realize midway, that we are suddenly more detached, less involved in the drama, more careful of where we are putting our energy and life force in order to make different choices. There is no fanfare. Nobody may pad our backs or notice, but we just have a feeling of knowing or a sense of peace inside that sneaks up on us, letting us know that we are now at a new level of our evolutionary journey.

What we "encounter" along our journey is really something within. Whatever happened on the outside was just a reflection which may help us realize our true inner priorities, the responsibilities we want to continue or discontinue etc. In this way time becomes our friend.

When we become parents for instance we can see that some of the traits we give our children are related to traits we carry with us from our parents who in turn carried this to us from our grandparents and so on. When we become middle aged we can see how some of the paths we have chosen in life relate to ideas or beliefs we have taken on from others without due consideration to their timely or personal relevance. For instance we may have been going full speed ahead into a known career or life direction where we sincerely thought we were going, but end up sidetracked or feeling like we are getting nowhere. Or we may have been going full love speed in a personal relationship, but suddenly find ourselves in an empty no-man's land – where our love "dries up". Where we are, how we got there or feel about it, and what we are going to do with it next are all important to our ability to exert life, so that we do not have to repeat or do the same thing the same way ever again.

If we suddenly find ourselves in a void between cycles - which is the case for many people on the planet right now because of the Shift, where many things are "in between" endings and new beginning or "up in the air", this time is altering our perception of where our life is going. It could be because our path is being altered, willingly or unwillingly because there is something we need to learn from something, somewhere else. Somehow, this lack of clarity might allow us to remove or eradicate calcified elements from old landscapes of life before we can move into a new direction. It is all about going through these cycles.

Any new direction can be anticipated or arrive like a wild card event. Even though we may not agree to see it at first, being open, willing and able to adapt to life's different circumstances and events, can greatly assist us with what life has in store for us next in this new millennium.

Things that have been with us for eons may suddenly Shift space or direction as they no longer exist in the ways we thought they would, as wanted or needed. This includes people, activities, places, habits and beliefs which may suddenly drop out of our lives without caveat or warning. Anything that keeps us tied to too old or expired roles, personas, limited concepts or ways of thinking, may have to go in ways we cannot anticipate at the end of a (time) cycle.

Unlike what we might think, this can actually be quite helpful because when or if that happens, it helps us rediscover how we can live life with greater authenticity, accuracy and trueness. Whether this is frightening, challenging or exciting greatly depends on our view of things or outlook.

For instance; if we have been living mostly on the "outside", chasing only the material ways or sides of life - it can be difficult to suddenly have to surrender the idea that accumulation of money and wealth is needed to find inner happiness. Similarly, if we have been living too much "on the inside" thinking the outside world is not that important to us, suddenly having to deal with practicalities or local world situations can be equally as frightening as they are important.

Although this of course is not explicit to the Shift, it is nevertheless relevant because such revisions of life provide whole new opportunities that are very much a part of changing a personal or collective cycle in order to find a new balance or direction. When old cycles end, there are things we cannot take with us, but there are also many new things that are made available to us when a new begins. We just need to figure it out.

*To everything there is a season,
a time for every purpose under the sun.
a time to be born and a time to die;
a time to plant and a time to pluck up that
which is planted;
a time to kill and a time to heal ...
a time to weep and a time to laugh;
a time to mourn and a time to dance ...
a time to embrace and a time to refrain
from embracing;
a time to lose and a time to seek;
a time to rend and a time to sew;
a time to keep silent and a time to speak;
a time to love and a time to hate;
a time for war and a time for peace.*

Ecclesiastes 3:1-8

PART 4

BIRTHING THE NEW REALITY

What is below is like what is above. And what is above is like what is below, so that the miracle of the One may be accomplished.

— Tabula Smaragdina

Photo (previous page)

Image credit: © Fotosearch

CHAPTER 9

A HIGHER WORKING CONSCIOUSNESS

Envisioning New Ways of Being & Living

The universally heralded and perceptual "Shift of the Ages" changes both our inner and outer landscapes – the micro and the macro – inner and outer world of our lives. In this sense we find ourselves "in between" as we pass from the old cycle or era and step into a new reality or age for the world.

The new paradigm is emerging as the Earth is adjusting to a new place in the cosmos and as we gradually awaken to a higher working consciousness on the next step of our human evolutionary journey. This is telling us to follow our real being, instincts and hunches more, while putting past events, emotions and thoughts into a whole new framework or perspective. This is more spiritually based as we learn to relate to our human interconnectedness and all life.

It is awakening the world mind and our human potential in ways that can take us beyond past assumptions, limitations, fears, doubts or restrictions that have prevented us from understanding who we really are and how we can envision new ways of being and living while creating a world that we truly want.

One of the essentials of stepping into this new world or reality is diving into the spiritual, unlimited, timeless aspects of our inner Self before we go out in the world and leap into meaningful or affirmative action that can actually address the outer landscapes of life – in ways that enlighten physicality.

It is not like we are going to sit still in a cave and meditate on how our spiritual Self relates to worldly reality. On the contrary, it is about bringing more of our one, true being into action by allowing the beauty of who we truly are as human beings to infuse our physical reality. It involves bridging the personal with the collective. We can do this by learning to work with the multidimensionality of reality and human intelligence in such a way that the One Creator, God, Creation - or whatever you want to call it - can better work through us or on our behalf to create more of "Heaven" here on Earth. Most of us, however, have bought into the opposite idea that we have to ask God or someone else for permission to create what we want or that Heaven is somewhere far away we go or escape to when we die. Hello! Is it any wonder then why we have sometimes felt so separated or disillusioned about being here - in the past cycle?

If we are to turn this notion or past around, we need to look closely at who we are and what is essential to us while assessing what we have to do in relation to what we want, is really needed or happening "out there".

For instance, if we have previously made (too many) choices or decisions based on misconceived notions about our potential or the outside reality stemming from ideas or beliefs that were handed down to us from our parents or grandparents, old "dusty" books, expired or derailed concepts, we need to start seeing and moving beyond these and make different realisations, decisions and choices that are based on the real power of truth. Past assumptions about reality and our human cap-abilities are probably not going to quite match up anyway, with who we are in the process of becoming, where we are going, or how we are going to live in the future of the next cycle. We need to start "shaping up", believe and behave in ways that allow us to do things differently in a new framework that are more in alignment with our power and the possibilities of an emerging new reality.

It is about finding or discovering the things that are really calling out to us, that make us feel inspired to go in new directions that better reflect who we are or are in the process of becoming as we become more connected to our inner self or life purpose. What we consider possible is often far from what we could do if only we changed our perception of reality and discarded much of what we have believed is true or considered rightful about life here on our planet in the past.

In the process, we may discover that this is something we have always known or felt deep inside which is now coming to the fore as part of us or the world around us is changing while we realise what we came here to do to make life a more real, meaningful or purposeful experience.

It can be the things that magic and dreams are made of, even when we cannot always see or experience it, while we are this world or a future reality, because it is in the process of being created by our thoughts, words and actions. With so many things still up in the air before they settle down or land in their rightful positions somewhere or sometime way beyond 2012, is it any wonder if some of us are sometimes mulling over when or whether we are going to get there...?

When or if we are in a change or "make over" process that could last many years or decades we must believe in and nurture some of these incentives. What is important is to do what we can do to make things more right or better, right now, so that we can take our next step with more future confidence, knowing that doing more of the right things instead of the wrong things might eventually take us there.

If we are unhappy about something in life or the world, it is important that we do not reject or ignore how we feel about it or project our fears about what might happen onto the world. Instead, we have to stay ultimately real and be realistic about what is going on in order to change whatever needs to change, accordingly.

While there is no need to throw the past out of the window or with the bathwater – we should remember that sometimes we really have to let go of things in order to more fully embrace new opportunities that are better aligned with the future as it will or we will in order to "lift" us, as well as others by listening to life's "signals". Life on Earth can be a progressive evolutionary journey, if we allow it to be. Even if we feel that we or the world might be going astray sometimes in this shifting process, it is not permanent, but because we are going through a changing, shared reality.

The New Era Is Not Asking Us to "Sacrifice" Our (True) Self

If we feel that we have lost our way, been misled or misguided by "old- world-ways" or imbalanced, masculine, material or overt rationality, human egos and the like, through whatever life derailing process, we can learn to balance it with new, more feminine, intuitive, holistic, mystical ways of being in order to open our view or a window to a higher, more evolved or balanced human mind set. By releasing what has occurred or impaired us in the past (in) life, we can find a higher position or state of being or living in consciousness which allow us to move forward in the pursuit of a direction that we believe will work or be better for us because it is more in alignment with who we are in relation to who we are becoming through our next steps...

The unlimited, timeless aspects of self carry the ability to heal old issues and ways of life we have been unhappy about. Remembering our Spirit in this way - every day - can invigorate us with new energy, insight and visions for a better life. As we get to know and learn to trust the spiritual or real Self more - without becoming pretentious or fanatically "religious" about it – the universe can respond and show us ways of being, acting and living which can guide us through life's evolutionary pathway and journey.

Humankind, in mass today, still does not recognize or understand their relationship with divinity in this way because the true or spiritual depths of self are often regulated to the realm of the ego personality by the masses. Therefore, we are only beginning to grasp our true human potential and spiritual possibilities as they become more available to (more of) us. That probably goes for you too.

The old ego-mind complicates everything and sets itself apart from higher aspects of consciousness because it so often feels threatened or separated from life in order to survive. The true, inner or authentic self, on the other hand, simplifies. It "knows" because it is animated by Spirit, so it is vital that we as human beings search to understand its meaning; how Spirit works through us and how the dynamics created by the ego opposition the energies of self which could otherwise collaborate with the Big Spirit.

The question of course is, whether or not we are tracking on this and whether the time beyond 2012 is going to be a breakthrough or breakdown path?

Most of us have been told to use our mind exclusively as an intellectual tool to direct life or control whatever reality, but intelligence is so much more than intellect. And being in control is so much more than always "taking" control. The human mind or consciousness includes the emotions and the feelings of the heart, our soul and the spiritual gifts as well. Moving away from the limitations of the ego mind involves an expanded or higher level awareness in a new spiritual context that allows us to use the multidimensionality of consciousness. This automatically brings forth the truths about the relation between ourselves and others and reality through interconnectedness. This is greatly needed If we are to co-create a different reality for ourselves and the world at large than the one which we have gotten often mindlessly used to during the previous cycle.

We have to get out of the traps of the old ways we have created if we are to embrace the new era, cycle or future that is coming towards us. It is not so much an analytical process as it is about merging our authentic self with reality through a greater awareness that includes our true feelings and our hearts as well. We all desire to be accepted, loved and respected for who we are with or without receiving other people's explicit attention, but apparently this is also why it has proved so difficult in the old cycle to figure it out. If we only desire love or attention for our egos to show off with and are unwilling to show love and respect to others – how can we expect to receive other than something minimal which only temporarily resembles love?

How many human lives or relationships mirror the lack of love we feel or the fears that prevent us from expressing who we are? Is there any doubt now about the consequences? If there is only fear or strife how can there be peace or love on our planet?

How can our human lives then come to reflect a natural deserve-ability in relation to our living and being here in this life?

The path of the more limitless, timeless Self starts with saying yes to our selves and to new ways of being that reflect it. Only then can we truly say yes to new ways of life that allow the world to become an outside reflective image of what we want to be like.

The kinds of emotional dramas that play themselves out - through our human relations an interactions stem from the fact that we are not ready to change all this or that yet. The world reflects our inner states. From then on the drama continues because we need others to confirm or establish that we are "lovable" through all kinds of fearful, manipulative tactics or methods where the ego craves this or wants that in order to feel good or ok in some way even if it destroys our societies or the Earth itself. The path to a higher level of evolution in consciousness for mankind goes through the unlimited self by transforming the lower emotional body through the heart. Until we are willing to look our ego in the eye and become aware of how it often distorts our reality, how can we expect to see what is really going on or needed within our world or human relationships?

The first thing to do is always to make an inner decision to change that which allows us to align with a different place in consciousness so that our outside world and relations can come to mirror this as well. It really all starts with(in) us.

177

Reaching for the timeless, unlimited aspects of self allows us to take a new, positive approach where we do not become trapped by the lack of authenticity or integrity of the human ego. We can either allow or disallow the inner Self to pave a new way.

An expanded human awareness or unfolding consciousness beyond its limited 3rd to 4th dimensional (physical-emotional) aspects can emerge into a new human form or template where we can actually realize our true spiritual potential beyond the old mentality of the ego mind. Most of us have experienced being with these higher states of consciousness at some point in life. It is not occult or mystical. For instance when we are engaged in love and open ourselves to our true (inner) nature. We all know what it feels like when we are with a beloved and the stars reach the Earth or they help us reach for them. This is just one important aspect of the open or multidimensional consciousness that stretches beyond the visible into the invisible that we are all apart of, but do not normally see with our human eyes. The invisible is an energy force field that penetrates all living things and the universe. Understanding this takes many physical forms and teaches us about our relation to all of Creation.

The Emerging Co-creation & Unity Consciousness

The new waves or energies that are arriving in steady incremental pieces are increasingly altering the experience of creation and consciousness in time and space.

For a while we might experience past and future through an intersection that allow us to see our past (mis-)creations in relation to possible (better) future scenarios if we can access the levels in consciousness where we become aware of the fact that we create our own reality. The more we become aware of it, the easier it becomes to create from a higher state of consciousness that transcends our past conditionings so that our authentic self is not rendered useless or we become "victims of fate" instead of (divine) co-creators of our destiny or reality. It is not an easy idea or concept to grasp if we are mostly familiar with the opposite.

For instance, in the past cycle we have gotten used to thinking that we "know better" or "what to do", but what if what we think we know or know how to do takes a wrong turn or turns out bad for us long term? Just look at the planet's degrading environment for instance. If we think we know it all in some linear fashion always or already, we may need to think again if our ego is doing all the talking or thinking and our inner Self has no saying over whatever decisions or scenarios. There is no way around it – we have to let go of the old separate ego mind - if we are to evolve beyond it and embody more unity or co-creation consciousness which allows our spirit to lead the way. It is not so much about creating what WE WANT ALL THE TIME than it is about what we need or what is GREATLY NEEDED in the group AT THIS TIME by sharing a more similar or creative vision, intention or purpose with for a common good. Only when we allow these higher aspects of self to shine through, can a higher vision or reality be created for the future of our planet.

In this new understanding or connection with a higher level of consciousness, our ability to co-create can be increased manifold by sharing with other people within the community or a group. For instance, if we can move our consciousness into a higher (5th) dimensional time/space point we would be allowed to create almost instantly because it embodies more love and connection to the higher creation field matrix. In many ways this alignment has not been possible before when many or most people have been wrapped up in their own selfish goo or limited thinking in linear time with now consideration for others, a greater or higher plan in mind.

In our (shared) physical reality, the ego mind's job is to help us survive, so it puts forth its own separate creational intentions, whereas in the forth-fifth dimensional reality, creation consciousness moves up or into the heart, based on higher emotional ideas that we have to help or service others and so on. Only when we fully embody the 5th dimensional reality of creation, where a sense of oneness, pure heart love or unity based thinking in consciousness takes over, do we become deeply aware of, enmeshed and entrenched in our interconnectedness with other human beings and all life and the unified field that surrounds us so that we can start to work, co-operate or co-create with the universe through an evolved, shared conscious focus without personal or hidden agendas getting in the way of collective ideals or higher objectives.

Initially, this may seem out of reach, far-fetched or like a utopia for many people, but when we consider the often dysfunctional way our societies or organizations operate even today, the more we come to understand this level of consciousness, the more we may want to explore it and revise old ways of creating and doing things.

If we want to expand on the benefits of shared co-creation rather and simple co-existence we need to look at how our creational abilities can match the context in order to be able to construct a new paradigm or reality in the future world we are going to live in. Higher levels or aspects of consciousness allow us to come into greater alignment with all of creation and the energy that creates and weaves and builds reality. If we can let some of our old ways go, the new ways of building or creating by sharing energies within the unity consciousness field of a group, can make creation faster, easier, stronger and more fun because there is less inertia or resistance when our co-creational ability is not "disturbed" by the lower 3rd-4th dimensional (creations of the) mental-emotional body. It basically means, we see things as part of the whole rather than as separated parts and that when we are together with other people we see their natural gifts that they can share with us like the ones we want to share with them. It is a co-operate way of engaging our spirits by sharing and giving of what we have to offer, instead of thinking of how we can take away from or control others, to feel or look good whole leaving other people drained. We can better (co)create if we come together instead of splitting ourselves apart.

Each of our individual consciousnesses holds a holographic key to co-creation because we are all here as a part of the whole or a piece in the Big Spirits "puzzle". So, when it comes to figuring out how we can create a better future for us – co-creational or unity consciousness is the way to go.

It is how we can learn to contribute to a new planetary matrix of creational possibility and magic in our local and global community, but of course it is going to take time before we fully understand the implications of these concepts that allow us to work together to create and share visions.

There is not one way or pathway to do that, but many because it is about learning to understand how each of us can deliver our unique contribution while allowing for the best option or highest vision to manifest. We cannot over-rationalize or intellectualize our way into it. We have already done that to death in the past. We have to learn to use open-source intelligence where integration of ideas, thoughts and reality principles combine to form the pieces of a grander context or puzzle.

For instance, an increasing number of people are seeing amazing results through "Cosmic Ordering". This involves giving the universe or cosmos a clear and concise request while continuing to do so until what you/we desire is granted. Of course, many people might still prefer or like to believe it to be complete hogwash, but if we appeal to the "all good things" come to those who wait if it is for the highest good of all concerned, adopting this philosophy might be nowhere as far off as we think. At least if or as long as the things we desire are truly close to our heart (of hearts).

From where the physical and the non-physical or spiritual realities meet, we can learn to understand how intentional energy follows thought and creates our reality. We can also trust that our souls know the way or how to unfold the focus of our intentional experience by realizing that our true spiritual self goes *BEYOND* earthly experience and exists through multiple focal points —some with form, some without. All we want to create (already) exists in this realm of the All-That-Is - if we can open our (higher) minds to it.

Co-creation comes with insight and responsibility. Only when we embody a higher state of consciousness which takes ultimate response-ability for the fruits of its wisdom can we align with a higher vision, where God or the Universe can assist us in growing the (co-)creative seeds that we are sewing for something other than just ourselves. Surrendering to co-creation with courage, integrity and humbleness allows us to align with the highest good. It is not that complicated – as a matter of fact the seeds of creation are already stored within you.

"WHEN MAN LIBERATES HIS INNER
POTENTIAL HE BECOMES POWERFUL
AND EVERYTHING IS POSSIBLE."

- BOLIVIAN SHAMAN, CHAMALU

CHAPTER 10

THE COSMIC AND CO-CREATIVE SELF

Each Soul Has Its Own Pathway

Sooner or later every man and woman wakes up to the fact that they are going to die one day or they are divine human beings. Despite the appearance that we are only physical in nature and that reality is filtered through the perceptions of our physical brains, there is much more to each human being than their physical aspect. It is still a minority, however, of the human population who is aware of, fully understanding or using it.

Above our physical brain is our mind, which is more of a field of (collective) consciousness. Within that field of consciousness, we exist as an aspect of the divine source from which we came through our physical, emotional and mental bodies.

Long ago, we inserted our selves into the human experience with our consciousness. At that point nothing was yet physical and being human meant being a specialized personality with the free will to explore consciousness in greater detail. Wow! A physical or incarnation experience was sought and it made being a rational, free will-driven, unique individual like you all the more interesting or exciting. The "lights" went out for a while and from that point on, your origin in the cosmos was replaced with a physical existence which meant living in the duality and density of physical life here on Earth, but what about your conscious awareness of your true self remembrance and its relation to the cosmos? Most of us have not got a clue or have entirely forgotten as we "grow up".

In the past time cycle or paradigm we have been immersed deeper than ever before in the human experience of being physical. It happened because we needed to explore our relationship with physical matter, fully to understand more of the light. Now that two Sun and time cycles overlap we find ourselves in a new field or intersection – similar to when two intertwining coloured circles mix on a palette. It shows us the colour of the past cycle in contrast to the next circle or cycle while showing us an interesting mixed colour in between while we are looking for the colours we want to use to paint the next picture. We look at it and wonder. Similarly, our awareness about self or soul or the spiritual colours of life are now coming back to the fore, because they are coming more into focus as two cycles' overlap.

We are not quite there with the old coloured cycle, seeing or feeling the new one yet. It is as if we are being squashed between two circled colours sometimes with lack of clarity as the different colours flow together and mix into our limited consciousness. The "in-between" time is awakening us once again to our true natural colours on life's palette before we can move on with the painting ahead. This is as interesting as it is important or challenging.

Being mostly disconnected from what we really wanted to paint in the old cycle meant that we drifted away from our awareness of what could be painted with more delightful colours. We have to earn our colourful realizations in this way through direct experience or dedication to our work as human beings as time cycles progress.

Our True Inner Nature

As humans, we have (self-) awareness, intellect, and freedom of choice within the mysterious maze or thrill ride of being in a life here on Earth. As we are wrapping up certain experiences of the past cyclical maze, there is a treasure to be found - the realisation that we are not lost or separated form the light of our souls or God, which have been there all the time, yet "hidden" from sight.

When we as mice look for the "cheese" in the maze, it begins to beckon us forward towards the end goal or into the realms of more light. Then, we are firmly on the path of our spiritual discovery, what some experience or call enlightenment. Some are heading back along that pathway now to the place they came from before the thrill ride began. Others are only starting to realise some of what it implies.

Remember, our brain is not our mind and our mind is not our personality. Our personality is not who we REALLY are. Our immortal soul is our "real personality" and it is functioning through the filter of who you/we became through the social conditionings and other lifetime experiences, you/we have had.

If we attune ourselves with our soul or inner consciousness and remember spirit every day, we will always be able to make things right or make more rightful choices based on who we truly are and where we come from and not from something that others or our ego minds intend to do or tell us to be. When you sense the essence of your true inner nature instead of your limited or lower self personality, you can sense which of the options that lie ahead before you resonate best with your inner core or guidance attuned as you are to the Self (with)in time. This is your real potential or perhaps secret to gaining the most from your life experience as a human being at this unusual time.

It is that "silent knowing" within you that is not telling you to do this or be that from "negative emotions". The Soul within or the "Voice of the Spirit" - is a "silent voice" that guides you when you ask the Big Spirit for help & clarity along your way.

What is happening, during this Shift, is that many people keep trying to do things "the old ways" – that comply only or mostly with their ego - even when they are not working properly or anymore. This creates a disturbance, unsettledness or confusion and what lingers on is a growing sense of restriction, dissatisfaction or inability to cope with life. It does not bring about the change, inner peace & resolution each of us would like to see, sense or experience as true, evolving or creative human beings that are heading towards a new future.

Our transitional years through 2012 and beyond mark the final beginning of human life where we learn to work with the Greater Plan or voice within as it was meant to be. It is like a graduation. We can accept and work with this curriculum to pass the test processes so that change becomes much easier or perhaps reject it. It may take something from us, but eventually it can bring enormous benefits and returns. This is why it can be one of the best and worst of times or perhaps most exciting time in human history to be alive. We are about to experience a transformation of the human Self and culture that our ancients used to only dream about. It is going to be or become a more spiritual one...which will look and feel different, but of course this will not appear or happen overnight.

As we awaken to our true spiritual potential it becomes like a snowball that begins its descent down a tall mountain. Initially, there is only a handful of snowflakes that gather bits of ice, then the ball grows bigger as it rolls passed tree branches and rocks, becoming ever larger and stronger. As it gains momentum, it is rolling down the mountain faster and faster till it becomes so large or travels so fast, that it cannot be stopped anymore and eventually blocks our (old) view.

This is just to give you an idea or image of our shared evolutionary journey's potential during this shift of cycles. The question is not if, but how much, when or where we are going to grow from the process. Although diving into new depths of self may seem like another universe at first for many people– it is simply a new key to our unfolding human self and evolution through time. It is like turning our hearts toward God and our minds toward a more spiritual consciousness. As information travels more rapidly and easily around the world than ever before now, we can even share the growth experience with like minded others.

Sharing Our Collective Soul Mind

The present Shift into more "spiritual" is also being catalyzed by the disillusionment of many people with old corporations, establishments or institutions that are governed by the manmade self-interest or greed of the ego-mind, when what we need is a new shared focus on responsibility, compassion, cooperation and unity of service that can take us to the next evolutionary level if we allow our human experience to align with our (shared) purpose.

In this way our individual soul path relates to the cosmic path and that of our collective body where we can benefit through greater awareness and knowledge of our true potential and nature – instead of identifying with the ego and its false beliefs. Following this path allows us to trust our selves from within and accurately trace or diagnose internal cause from observable, external effects in our shared reality. It is only through this greater awareness of self that we can fully comprehend the universal language of co-creative possibilities that are being made available to us.

In a way, we need this "spiritual re-education program" because most of the education or training we have received in life often comes from the limited views of the intellectual mind or are based on ideas that have reached their expiration dates or become ignorant to the spiritual truths of life. This has made us overlook the innate human intelligence of the soul and separated us from each other, struggling to survive while destroying the beautiful world or planet of which we are all a part.

When we search within we can better find the answers to our or life's deepest questions, what we deeply and truly feel we need or need to do and how to find it as required. Nobody can tell us what is right or wrong, because the truth about this should be drawn from the soul within or the Higher Mind.

Opening the Higher Mind

As we continue through the Shift what is essential might come more to mind or the fore as we are beginning to learn how to access new levels of consciousness where the Soul and spirit interact through the Higher Mind. It is through this Infinite or Higher Intelligence source that a new version of being in a human form or way of life becomes accessible in a more conscious manner that is not restricted by linear thinking or distortion through pure analysis or intellectual logic. It can arrive in increments like a wave or new structure of awareness, which creates a deeper understanding or instantaneous knowing through a single, quiet, clearly focused point of observation.

It is through this Higher Mind Source, the Creator of all that is can become known. In the ancient Sanskrit language this connection is referred to as the "Antakarana," – the rainbow bridge - or return to God. This can be built through a combination of soul-communication, prayer and reconnection with God through the 90 percent portion of the brain or DNA that science believes is redundant, does not function or exist. Even Einstein, when he proposed the universal theory of relativity, believed he might only have used 10 percent of the brain, so what about the rest of it? If we are beginning to understand or merge with our Higher Mind, does it mean awakening to spiritual insight, inspiration or psychic gifts that allow us to use a higher vision of what life can be like on Earth?

> *"I want to know the thoughts of God, all the rest is just details."*
>
> *Albert Einstein*

Infinite Intelligence as Source

If we can disengage from the mind as pure intellect and logic and come to understand the higher mind as intelligent consciousness, we can learn to access new instant or intuitive knowledge, through the truth and wisdom of our soul's inner knowing or yearning. Sometimes when we are in quiet focus, meditation and detachment, we can experience a new sense of mindfulness or inquisitiveness without asking a question to find new answers to problems or challenges, we are facing globally or personally.

Because such opportunities are becoming increasingly available to more and more people, bridging or merging the left side of the brain's rationality and logic with the intuitive, non-linear or creative aspects of the right side of the brain. Unity or multidimensional consciousness is something we can all learn. If this is our destiny through 2012 and beyond our true abilities can become available to us once again so that we can begin to replace old limited ways of being, seeing or doing things through a more timely awareness and deeper understanding of what we need to do.

Merlin's Crystal Ball – Developing a Clear Vision

When we move through the end point of time – time can stop or open up like a jewel box - revealing past, present or future scenarios. What once was (going to happen) could be changed or altered in an instant – with our thinking or perception about it as we and it is passing through a zero point. By allowing ourselves to see through the illusions of time through the past, present and future - the timeline(s) we are travelling on could become clear or make themselves "known" as if we were gazing through a looking glass or staring into a crystal ball. In this way, we might come to see or experience different realities or dimensions including potential future(s) and their outcomes or the past in a whole new light. It depends on the visionary and other capabilities of our expanding minds.

The present, as the word says, is indeed a gift. Perhaps because it makes past and future time more accessible through the Now?

What if we, without experiencing the emotions attached to it, could observe the exact moment in the past when threads of disharmony were created? What if we could alter that experience by substituting it with a different reality that demonstrates a more masterful future choice?

What if we could return suppressed skills and knowledge to access our highest human qualities and integrate them with new conscious choices now or in the future - all because of an eternal NOW experience?

Even if this seems farfetched, the Shift might eventually show us through direct experience what a timely reality really mean as we spiral through the end of the old cycle and into a new one...

Resetting Time through the Looking Glass

If we can clear the past through the present and visualize a new reality, it seems possible to clear our future path ahead because what is no longer determined by the past – does not precondition the future in the same way it has or did in the past. Perhaps we can better create the future, if we lay our experiences of the past or old time cycle to rest.

Isn't it an interesting thought then that if we create (our) reality, we might be able to step in and out of time and reality? If time becomes more elastic or plastic almost like a surreal Dali painting, the way or what we chose to think or do around or about it may not just affect the present, but also pop up on a future time screen. Hopefully we are wise enough to understand the implications of what it means to project time and realities if they open up like a jewel box. Just think about it for a moment if you could reset time and reality. Would you know what to do about the future? What visions would you have if your visions could become a future reality?

Choosing the Future World Beyond 2012

Even though the time frame beyond 2012 is full of uncertain future events or possibilities because time and reality might be "up in the air" or changing as we go along, it is important that we allow ourselves to understand how we can envision a future reality that better fit in with the "Greater Plan".

Our ability to impact (future) timelines always occurs right now because Real Time is a simultaneous time where Creation is concerned because nothing is really totally separated. Only our human understanding of time and reality through the measurement of linear time allows for a division of timely realities.

In order to experience Simultaneous or Real Time, one must become consciously aware of time and the mind's limited linear thinking. An expanded consciousness could "blow our minds" outside a third dimensional linear, physical reality.

Consider for a moment the possibility that you could step outside of normal time and rearrange any event you have experienced or are about to experience by clearing the disharmonic threads and replacing them with the wisdom and possibility of a new outcome for a better tomorrow. ...what would you do about the future? Would you have the wisdom, faith and courage to take the steps or actions? Would we only like to see what we see in a meditative or a dream state with or without the nightmares?

Indeed our planetary focus and future depends on our state of consciousness. If we truly understand that we have a rendezvous with our future (destinies) through 2012 and beyond, it becomes both our opportunity and responsibility to choose wisely without bringing to many fears or bad dreams into the picture. After all, what would we do or what would happen if our unconscious wishes or visions suddenly were made manifest or appeared more quickly? Would we have the knowledge and wisdom to merge personal insight with a higher collective vision? If more or all of us did - that would probably change the future of our world in an instant........

No matter how we look at it or what is going to happen...

This is perhaps why the Shift of the Ages through 2012 and beyond remains the biggest change challenge of our time....

Are you ready?

A Vision Quest

The Shift of the Ages is not the end of the world.

It is going to be the beginning of a new one.

It depends on us & our ability to (re)create it.

Merging & converging timelines

How would we want it to be?

Can we imagine what it would look like?

Can we make our vision(s) happen?

Where? When? How?

References

José Arguelles: The Transformative Vision

Jose José Arguelles: The Mayan Factor: Path Beyond Technology

Carl Johan Calleman: The Mayan Calendar and the Transformation of Consciousness

John Major Jenkins: Maya Cosmogenesis 2012: The True Meaning of the Maya Calendar End Date

Terence McKenna: The Invisible Landscape: Mind Hallucinogens and the I Ching

Sylvanus G. Morley: The Ancient Maya

Daniel Pinchbeck: 2012: The Return of the Quetzalcoatl

Made in the USA
Charleston, SC
05 April 2012